2018/2019
THE STATE OF CHINA'S CITIES

Global Action & China Practice:
Together for a Better Future

EDITED BY
China Science Center of International Eurasian Academy of Sciences
China Association of Mayors
Urban Planning Society of China

CHINA ARCHITECTURE & BUILDING PRESS

审图号：GS（2020）2613号
图书在版编目(CIP)数据

中国城市状况报告 2018/2019：全球行动与中国实践：共创人类美好未来 = THE STATE OF CHINA'S CITIES 2018/2019 Global Action & China Practice: Together for a Better Future：英文 / 国际欧亚科学院中国科学中心，中国市长协会，中国城市规划学会编著 . —北京：中国建筑工业出版社，2020.1
ISBN 978-7-112-24814-8

Ⅰ.①中⋯ Ⅱ.①国⋯②中⋯③中⋯ Ⅲ.①城市建设—研究报告—中国—2018-2019—英文 Ⅳ.① TU984.2

中国版本图书馆CIP数据核字（2020）第022550号

书籍设计：付金红 李永晶
责任编辑：杨 虹 牟琳琳 尤凯曦
责任校对：王 瑞

THE STATE OF CHINA'S CITIES 2018/2019
Global Action & China Practice: Together for a Better Future
EDITED BY
China Science Center of International Eurasian Academy of Sciences
China Association of Mayors
Urban Planning Society of China
*
中国建筑工业出版社出版、发行（北京海淀三里河路9号）
各地新华书店、建筑书店经销
北京雅盈中佳图文设计公司制版
北京雅昌艺术印刷有限公司印刷
*
开本：880×1230毫米 1/16 印张：12 字数：291千字
2020年1月第一版 2020年1月第一次印刷
定价：108.00元
ISBN 978-7-112-24814-8
（35378）

版权所有 翻印必究
如有印装质量问题，可寄本社退换
（邮政编码100037）

EDITED BY

China Science Center of International Eurasian Academy of Sciences

China Association of Mayors

Urban Planning Society of China

ACKNOWLEDGEMENTS

Hereby we specifically thank China Architecture & Building Press and its production editors, the translation service provider (Beijing Leader International Consulting Service Co., Ltd.) and proofreading experts. We also highly appreciate the support of School of Architecture, Tsinghua University, China Academy of Urban Planning and Design, and relevant departments of the Ministry of Housing and Urban-Rural Development in preparing this report.

EDITOR-IN-CHIEF

Wang Guangtao, Vice President, China Science Center of International Eurasian Academy of Sciences

HONORARY EDITOR-IN-CHIEF

Huang Yan, Vice Minister of Housing and Urban-rural Development of The People's Republic of China

EXECUTIVE EDITOR-IN-CHIEF

Mao Qizhi, Academician, International Eurasian Academy of Sciences, Professor, School of Architecture, Tsinghua University

Shao Yisheng, Academician, International Eurasian Academy of Sciences, Research Fellow, China Academy of Urban Planning and Design

Shi Nan, Executive Vice Chairman, Secretary-General, Professorate Senior Engineer, Urban Planning Society of China

COORDINATORS

Liu Honghai, Secretary-General, China Science Center of International Eurasian Academy of Sciences

Wang Changyuan, Secretary-General, China Association of Mayors

Qu Changhong, Deputy Secretary-General, Senior Engineer, Urban Planning Society of China

Yang Rong, Inter-Regional Advisor, United Nations Human Settlements Programme (UN-HABITAT)

Zhang Zhenshan, Programme Manager for China, UN-HABITAT

AUTHOR'S TEAM

Mao Qizhi, Academician, International Eurasian Academy of Sciences, Professor, School of Architecture, Tsinghua University

Shao Yisheng, Academician, International Eurasian Academy of Sciences, Research Fellow, China Academy of Urban Planning and Design

Shi Nan, Executive Vice Chairman, Secretary-General, Professorate Senior Engineer, Urban Planning Society of China

Yin Zhi, Vice Chairman, Urban Planning Society of China, Executive Vice President, Professor, Institute for China Sustainable Urbanization Research, Tsinghua University

Lin Jian, Dean, Professor, Department of Urban and Regional Planning, College of Urban and Environmental Sciences, Peking University

Zhang Quan, Vice Chairman, Research Fellow-level Senior Engineer, Urban Planning Society of China

Shi Weiliang, Vice Chairman, Urban Planning Society of China, President, Professorate Senior Engineer, Beijing Municipal Institute of City Planning & Design

Zhang Shangwu, Director, Urban Planning Society of China, Vice Dean, Professor, College of Architectural and Urban Planning, Tongji University

Qu Changhong, Deputy Secretary-General, Senior Engineer, Urban Planning Society of China

Lu Qingqiang, Deputy Chief Planner, Director of Research Center for Master Planning, Senior Engineer, Beijing Tsinghua Tongheng Urban Planning and Design Institute

Liu Shiyi, Postdoctoral Research Fellow, College of Urban and Environmental Sciences, Peking University

Zhang Zhiguo, Vice President, Research Associate, Urban & Rural Water Research Institute, China Academy of Urban Planning and Design

Ye Xingping, Deputy Chief Engineer, Research Fellow-level Senior Engineer, Urbanization and Urban Rural Planning Research Center of Jiangsu

Shi Xiaodong, President, Professorate Senior Engineer, Beijing Municipal Institute of City Planning & Design

Xi Hui, Senior Researcher, Senior Engineer, China Rural Planning and Development Research Center, Shanghai Tongji Urban Planning and Design Institute Co., Ltd.

Zhang Guobiao, Senior Planner, Urban Planning Society of China

Foreword

Wang Guangtao

Vice President, China Science Center of International Eurasian Academy of Sciences (CSC-IEAS)
Member of the Standing Committee and Chairman of the Environment Protection and Resources Conservation Committee of the Eleventh National People's Congress, China
Former Minister of the Ministry of Construction, China

City is the common home of human beings, and bright future of the city needs our creation together.

From 2018 to 2019, with the 40th anniversary of reform and opening up and the 70th anniversary of the people's Republic of China as the time node, that China have gone through a few decades in developed countries after centuries of industrialization and urbanization road, creating Chinese miracle.

From 2018 to 2019, China's gross domestic product (GDP) has grown from over 90 trillion yuan to nearly 100 trillion yuan, and the per capita GDP will reach a new level of 10000 US dollars. The unemployment rate of urban survey is stable at a low level of about 5%. More than 10 million rural poor people have been lifted out of poverty. With the rapid development of urban and rural public service facilities and the cause of benefiting people, the living environment and standards of residents have been improved continuously.

The spatial structure of China's economic development is undergoing profound changes. Taking Beijing and Tianjin as the center to lead the development of Beijing-Tianjin-Hebei city cluster, promote the Xiong'an New Area, and drive the coordinated development of Bohai Rim. Taking Shanghai as the center to lead the development of the Yangtze River Delta urban agglomeration, focusing on the overall protection and non-development, relying on the Yangtze River golden waterway and giving priority to ecology, we will promote the coordinated development of the upper, middle and lower reaches of the Yangtze River and the high-quality development of the riverside areas. With Hong Kong, Macao, Guangzhou and Shenzhen as the center to lead the construction of Guangdong-Hong Kong-Macao Bay Area and drive the innovation and green development of the Pearl River-Xijiang River Economic Belt. Focusing on Chongqing, Chengdu, Wuhan, Zhengzhou, Xi'an, etc., it leads the development of Chengdu-Chongqing, the middle reaches of the Yangtze River, the Central Plains, Guanzhong Plain and other city clusters, and promotes the ecological protection and high-quality development of the Yellow River Basin. The comprehensive carrying capacity of central cities and city clusters has been continuously improved, and the functions of megacities and other densely populated economic areas have been orderly decentralized, and the development mode of effective treating of "city diseases" has been continuously optimized. With the construction of "one belt and one road", we will promote the coordinated opening of the coastal, inland and border areas, strengthen the interconnection of major infrastructures with the framework of the international economic cooperation corridor, and build a new pattern of coordinating regional and international development in China.

China's urbanization is in a critical stage. The level of urbanization has exceeded 60%. The main focus in Chinese society has been transformed into the contradiction between the people's growing needs for a better life and the unbalanced and inadequate development. Socialism with Chinese characteristics has entered a new era. The country will build a moderately prosperous society in an all-round way after solving the problem of food and clothing for 1.4 billion people and realizing a moderately prosperous society in general. The people's pursuit of a better life is becoming more and more extensive, which not only puts forward higher requirements for material and cultural life, but also increases requirements in democracy, rule of law, fairness, justice, security, environment and others. This is a new historical orientation of national development, and many problems need to be actively explored in practice. "People oriented new urbanization" will be an important development direction.

The State of China's Cities, which is jointly edited by the China Science Center of the International Eurasian Academy of Sciences, the China Association of mayors and the Urban Planning Society of China, includes Chinese and English versions, has been published since 2010, and exerting great influence at home and abroad. *The State of China's Cities 2018/2019*, with the theme of "Global action and China practice: together for a better future", echoes the theme of "Urban opportunity: connecting culture and innovation" of the 10th World Urban Forum to be held in Abu Dhabi, United Arab Emirates, in 2020, reflecting the common concern for human care and innovative development. It is hoped that the international community will work together to meet the challenges in the process of global urbanization, scientifically plan the blueprint of urban and rural development, exchange and learn from each other, and explore the road of sustainable development of cities in line with their own characteristics.

It is hoped that the publication of *The State of China's Cities 2018/2019* will help the international community to have an objective and comprehensive understanding of China's urbanization. Adhering to the urban work should take the creation of good living environment as the central goal, strive to build the city into a beautiful home where people and nature live in harmony, and continue to make due contributions to the sustainable development of cities around the world.

January 2020

Abstract

Shi Nan

Executive Vice President and Secretary General of Urban Planning Society of China (UPSC), Senior Planner

The term "transition" probably offers the most appropriate description of the current China's cities: the world's most populous country is entering the "urban era". The urbanization rate of China's permanent population has increased from 17.9% at the beginning of Reform and Opening Up in 1978 to 60.6% in 2019. More than half of the population lives in urban nowadays. In this huge change involving the migration of population, alternation of the social structure, and the renewal of urban and rural living environment, China has ensured 40 years of rapid economic growth while lifting 740 million Chinese citizens out of poverty. These outstanding achievements have never been made by any other country in the world at the same speed or scale. In addressing the common challenges of urbanization, China's cities have adopted many innovative explorations and attempts, which highlights the unprecedented significance of the choice of "Global Action and China Practice" as the theme of *The State of China's Cities 2018/2019* (hereinafter referred to as the "Report").

The ultra-large-scale high-speed urbanization has prospered China's economy, triggered urban transformation, also brought unprecedented challenges. In recent years, topics such as the prevention and control of water, soil, atmospheric pollution, the supporting security measures for migrant workers in cities, the protection and inheritance of historical culture, and the management of housing prices, have received widespread attention from the general public in China. The central government of China has responded to these topics with positive feedback. The report to the 19th National Congress of the Communist Party of China issued in 2017 proposed that the principal contradiction facing Chinese society in the new period has evolved to become the problem of unbalanced and inadequate development, and China will pay more attention to the issues of balanced development and fair development and will take a series of specific actions in preventing and defusing major risks, targeted poverty alleviation and pollution prevention and control, among others, to meet the current challenges.

In early 2018, China undertook drastic institutional reforms and promulgated a series of important policies intensively within the two years of 2018 and 2019 to make major adjustments to the national governance system, aiming to seek a balanced development model for economic development and resource consumption through comprehensively deepening our commitment to reforms, and transform China's existing development path. Among them, the reform of the spatial planning system is particularly important as planning spearheads China's urban development. As China's planning concepts and practices have always kept pace with the times and ensured the stable development of cities for decades, serious urban problems and large-scale urban decline have never appeared in China; instead, the quality of life of urban residents has continued to improve. It is foreseeable that in the future, China's planning will pay more attention to regional balance, urban-rural balance, and spatial balance, enable all citizens to access the same public services and the same quality of life, live in decency and dignity, and meet the people's pursuit of better quality of life. China's various practices will be reflected in this book, and we are also open to share and communicate with the world with an open attitude.

The preparation of this Report coincides with the start of the global implementation of the *New Urban Agenda*. Therefore, the Report puts an emphasis on the global perspective: it analyzes the common challenges facing global cities, actively integrates into the international context, and describes China's practical explorations in these fields. The Report is divided into six chapters, which are the urbanization process in China, spatial planning and urban governance, urban infrastructure, ecological civilization and urban environment, culture city, rural revitalization, and rural living environment.

The Report starts with an introduction to the top-level design of the future development of Chinese cities, i.e. the overall plan for the reform of the ecological civilization system, and offers an objective description of fundamental information on China's urbanization level, quality, and spatial pattern, etc., as well as a review of China's various measures aimed at continuously deepening the commitment to reform in such fields as land system, housing security system, The population policy and household registration system, urban investment and financing systems, and public services. The second chapter focuses on the major transformations of China's spatial planning and urban governance and gives the readers a panoramic view of the basic situation of China's urban development. In terms of urban infrastructure construction, the Report elaborates on the situation of urban transportation systems, water systems, energy systems, and environmental sanitation systems. In terms of urban environment governance, the Report introduces China's environmental governance in various fields such as the atmosphere, water, and soil. In terms of urban culture, the Report introduces China's practices in historical and cultural heritage, people-oriented urban design, and industrial heritage protection. At the end of the Report, five parts introduced the rural rejuvenation and rural human settlements environment in China, including rural revitalization strategy, improvement of rural living environment, poverty alleviation, small cities and towns construction and characteristic towns.

It is noteworthy that at the end of each chapter of the Report, we have selected the most representative cases of Chinese practice in recent years, which are diverse in type, rich in content, prominent in features, and highly valuable for reference. The appendixes to the Report are the industrial heritage protection lists published by China in the past two years, and the basic data of all 297 prefecture-level cities in China, including land, population, area of built-up district, economic indicators, and urban development-related indicators, which are one of the most authoritative urban development database for China.

With the use of a large number of data and cases, this Report depicts various aspects of China's urban development from policy guidance to project implementation, from urban and rural dilemmas to improvement measures, from international integration to local practice, and aims to enhance readers' understanding of Chinese cities, improve our consensus and contribute to the due strength of Chinese cities in order to make cities around the world more livable and sustainable.

CONTENTS

Chapter 1 Urbanization Process in China

1.1 Top-down Design of Urbanization / 003

1.1.1 Xi Jinping's Thought on Socialism with Chinese Characteristics for a New Era Becomes Guide to China's Urban Development / 003

1.1.2 Overall Plan for Reform of Ecological Civilization System Becomes Top-down Design of China's Urban Development / 003

1.1.3 Adhering to New Urbanization as Important as Rural Revitalization / 003

1.2 Urbanization Level and Quality / 005

1.2.1 China's Urbanization Rate Reaches Average Level of Middle-income Countries / 005

1.2.2 Living Quality and Standards of Urban and Rural Residents in China Improves Significantly / 005

1.2.3 China Adopted Human-centered Transformation Practices to Improve Livability Level of Cities / 006

1.3 Overall Pattern of Urbanization / 008

1.3.1 Influence of Chinese Cities in the World Urban System Continues to Rise / 008

1.3.2 Three Major Urban Agglomerations Lead with Regional Cooperative Development Promoted In-depth / 008

1.3.3 Metropolitan Areas Become New Entry Point for Improving Quality and Upgrading China's Urbanization / 012

1.4 Approach to Urbanization Development / 014

1.4.1 Development of Secondary and Tertiary Industries Creates More Job Opportunities; Development of Industry-city Integration Greatly Promoted / 014

1.4.2 Innovation and Entrepreneurship Promotes Employment Growth; Opening to the Outside World Enhances Development Momentum / 015

1.4.3 Explorations in Green Development Go Deeper; Green Ways of Working and Living Gradually Popularized / 017

1.5 Urbanization Supporting Systems Reformed / 018

1.5.1 Full Implementation of Residence Permit System Promotes Settlement of Floating Population in Urban Areas / 018

1.5.2 Reform of Three Types of Lands in Rural Areas Grants Farmers More Property Rights / 018

1.5.3 Housing Security System Improves to Meet Housing Needs of Low-and Middle-income Families / 019

Chapter 2　Spatial Planning and Urban Governance

2.1　The Process of Spatial Planning Reform　/ 023

2.1.1　Reasons for Reform: Fragmentation of Spatial Governance　/ 023

2.1.2　Exploration of Reform: the Practice of Multiple Plans Integration　/ 023

2.1.3　Direction of Reform: Unified Utilization Control　/ 025

2.2　Territorial Spatial Planning System　/ 026

2.2.1　Overall framework of the territorial spatial planning system　/ 026

2.2.2　Objectives of Building the Territorial Spatial Planning System　/ 026

2.2.3　Governance Improvement of Territorial Spatial Planning System　/ 027

2.3　Urban Governance　/ 028

2.3.1　Public Participation in Urban Governance　/ 028

2.3.2　Grassroots Community Governance　/ 029

2.3.3　Smart City Governance　/ 031

2.4　China Practice: Redevelopment of Underused Urban Land　/ 033

2.4.1　Concept of Underused Urban Land Redevelopment　/ 033

2.4.2　Effectiveness of underused urban land redevelopment　/ 034

2.4.3　Patterns of underused urban land redevelopment　/ 035

Chapter 3　Urban Infrastructure

3.1	Relevant National Plans and Policies　/ 041	3.3.3	Drainage and Local Flooding Prevention and Control　/ 052
3.1.1	Comprehensive Plan for Municipal Infrastructure　/ 041	3.4	Urban Energy System　/ 053
3.1.2	Plan for Sewage Treatment and Recycling Facilities　/ 041	3.4.1	Heating Supply　/ 053
3.1.3	Plan for Environmental Sanitation Facilities　/ 043	3.4.2	Gas　/ 054
3.1.4	Planning for Transport Facilities　/ 043	3.5	Urban Sanitation System　/ 055
3.1.5	Plan for Communication Facilities　/ 043	3.5.1	Municipal Solid Waste (MSW) Treatment Facilities　/ 055
3.1.6	Renovation of Old Communities　/ 044	3.5.2	MSW Sorting Pilot Project　/ 057
3.2	Urban Transport System　/ 044	3.6	Urban Communication System　/ 058
3.2.1	Regional Transport　/ 044	3.6.1	Overall Situation　/ 058
3.2.2	Public Transport　/ 046	3.6.2	Boosting broadband speeds and lowering rates for internet services　/ 058
3.2.3	Shared Mobility　/ 048	3.6.3	5G Technology　/ 059
3.2.4	New Energy Transportation　/ 048	3.7	China Practice: Smart City　/ 060
3.3	Urban Water System　/ 049	3.7.1	National Smart City Pilot Project　/ 060
3.3.1	Water Supply Security　/ 049	3.7.2	Local Practice Cases　/ 060
3.3.2	Sewage collection and treatment　/ 051		

Chapter 4 Ecological Civilization and the Urban Environment

4.1　Ecological progress　/ 065

4.1.1　Status of Ecological Progress in the Past 40 Years of Reform and Opening Up　/ 065

4.1.2　Characteristics of Ecological Progress in the New Era: the Report of the 19th CPC National Congress, and the Concept that Lucid Waters and Lush Mountains are Invaluable Assets　/ 067

4.1.3　Ecological Environment Improvement and Management: National Park System, and National Environmental Protection Inspection　/ 067

4.1.4　Ecological Security: Biodiversity, and Vegetation Protection　/ 071

4.1.5　Resource Security: Resource Protection and Energy Utilization　/ 072

4.2　Optimization of Ambient Air Quality　/ 074

4.2.1　Overall Condition of the Atmospheric Environment　/ 074

4.2.2　Distribution and Characteristics of Haze　/ 075

4.2.3　Haze Management—Implementation of National 10-chapter Air Pollution Prevention and Control Action Plan and the Blue Sky Defense War　/ 076

4.3　Optimization of Water Environment Quality　/ 076

4.3.1　Overall Status of Water Environment Quality　/ 076

4.3.2　Major Water Pollution Accidents and Pollution Control and Treatment　/ 078

4.3.3　Water Environment Control: Target Requirements of the 10-Chapter Water Pollution Prevention and Control Plan, River Chief System　/ 079

4.3.4　Urban water ecological environment construction　/ 079

4.4　Soil Environmental Quality Remediation　/ 081

4.4.1　Overall status of soil environmental quality　/ 081

4.4.2　Soil pollution accidents and pollution control treatment　/ 082

4.4.3　Improvement of Soil Environment: Target Requirements of the 10-Chapter Soil Pollution Action Plan　/ 083

4.5　China Program: Ecological Restoration and City Betterment　/ 083

4.5.1　Local practice　/ 083

4.5.2　Green Eco-district　/ 086

Chapter 5 Culture City

5.1　Cultural Heritage Deeply Rooted in History　/ 091

5.1.1　Overall Protection and Passing Down Civilization　/ 091

5.1.2　Upholding Ancient and Modern Glory and the Continuation of Context　/ 096

5.2　People-oriented Public Space　/ 099

5.2.1　Comprehensive Management of Street Space　/ 099

5.2.2　Open Regeneration Waterfront Space　/ 100

5.2.3　Space-building for Specific Population Groups　/ 102

5.2.4　Activation and Utilization of Vacant Spaces　/ 105

5.3　China Program: Industrial Heritage Protection and Reuse　/ 107

5.3.1　National Measures and Lists Promulgated: China Industrial Heritage Protection List (First Batch & Second Batch)　/ 107

5.3.2　Local Practice: Overall Protection and Renewal of Beijing Shougang Industrial Heritage Sites　/ 109

5.3.3　Local Practice: Comprehensive Protection and Development of Modern Ceramic Industrial Heritage in Jingdezhen　/ 110

5.3.4　Local Practice: Reopening and Utilization of Columbia Circle, Shanghai　/ 112

Chapter 6　Rural Revitalization and Poverty Alleviation

6.1　Rural Revitalization Strategy　/ 117

6.1.1　Rural Revitalization Strategy and Implementation Path　/ 117

6.1.2　Policy Promotes the Integrated Development of Urban and Rural Areas　/ 119

6.1.3　Planning Leads the Rural Revitalization and Development　/ 120

6.2　Poverty Alleviation　/ 123

6.2.1　From Poverty Alleviation Through Regional Development to Targeted Poverty Alleviation　/ 123

6.2.2　National Poverty Alleviation Plan　/ 125

6.2.3　Targeted Poverty Alleviation　/ 126

6.3　Improving Rural Living Environment　/ 126

6.3.1　Improving Rural Housing Conditions　/ 129

6.3.2　Improving Rural Infrastructure Levels　/ 130

6.3.3　Protecting Rural Historical and Cultural Heritage　/ 133

6.4　Small Cities and Towns Construction and Characteristic Development　/ 136

6.4.1　Development History and Roles of Small Cities and Towns　/ 136

6.4.2　Characteristic Development of Small Cities and Towns　/ 138

6.5　China's Rural Practice　141

6.5.1　Beautiful Villages: Local Practice for Improvement of Human Settlement Environment　141

6.5.2　Industrial Revitalization: New Practice of Comprehensive Development of Rural Industry　144

6.5.3　Cultural Protection: The Practice of Social Forces Intervening in Rural Heritage Protection　147

6.5.4　Joint Creation: Rural Governance and Construction　149

Appendix I　China Industrial Heritage Protection List　/ 152

Appendix II　Basic Data of China's 297 Cities at and above Perfecture Level (2016)　/ 158
　　　　　　Notes to the Basic Data of China's 297 Cities at and above Prefecture Level (2016)　/ 168

Chapter 1

Urbanization Process in China

Top-down Design of Urbanization

Urbanization Level and Quality

Overall Pattern of Urbanization

Approach to Urbanization Development

Urbanization Supporting Systems Reformed

>> 1

Urbanization Process in China

Since the founding of the People's Republic of China 70 years ago, China has promoted the largest urbanization process in human history and profoundly changed the global landscape. At the same time, many outstanding contradictions and problems have emerged during this rapid urbanization. Since its urbanization rate exceeded the world average in 2013, China's urbanization has entered a period of adjustment and transformation.

In order to meet the challenges presented by urbanization, China has followed sustainable development concepts and goals advocated by the international community, and has regarded thoughts of ecological civilization as the guiding ideology and action model for urban and rural development. It has further strengthened the people-centered governance concept; and established new development concepts in innovation, coordination, ecology, openness, and sharing; has transformed urban development patterns, optimized the overall pattern of urbanization, and also adhered to the 'two-wheel drive engine' approach, i.e. rural revitalization and urbanization while simultaneously reconstructing the governance system for territorial space.

In this process, China has actively explored the key tasks and themes, including the mutual promotion of urbanization and industrial development, the development model of green urbanization, the integration of innovative open platforms into global urban networks, and the convergent urban and rural development. Among others, "green, open, innovative and convergent development, humanistic care, and high quality" have all become new key words for urbanization development in China.

While respecting the law of urbanization development, the Chinese government has adhered to the ruling style of "bold actions, reform and innovation", and conducted in-depth explorations on urbanization supporting system reform as well as local pilot practice in response to real life problems and institutional bottlenecks.

The practice and explorations of China's urbanization will remain an important part of the world's urbanization process, and will continue to contribute to the wisdom that China and its programs can offer the world.

1.1 Top-down Design of Urbanization

1.1.1 Xi Jinping's Thought on Socialism with Chinese Characteristics for a New Era Becomes Guide to China's Urban Development

"Xi Jinping's Thought on Socialism with Chinese Characteristics for a New Era" proposes to modernize China's system and capacity for governance according to the five-sphere integrated plan and to build China into a great modern socialist country that is prosperous, strong, democratic, culturally advanced, harmonious, and beautiful by the middle of the century, thus indicating the direction for urbanization development in China.

The principal contradiction facing Chinese society in the new era is between unbalanced or inadequate development and the people's ever-growing needs for a better life. China must therefore continue its commitment to the people-centered philosophy of development, and endeavor to promote well-rounded human development and common prosperity for everyone. China's economy has been transitioning from a phase of rapid growth to a stage of high-quality development, and China has upheld its five-concept vision of innovative, coordinated, green-minded, open, and shared development to guide urbanization.

1.1.2 Overall Plan for Reform of Ecological Civilization System Becomes Top-down Design of China's Urban Development

The development of an ecological civilization is of fundamental importance for the sustainable development of the Chinese nation. It is imperative to adhere to the fundamental policy of harmonious coexistence between man and nature, with the development concept that lucid waters and lush mountains are invaluable assets, the guiding principle that a sound ecological environment provides the most inclusive benefits to people's wellbeing, and the systematic thinking that mountains, rivers, forests, farmlands, lakes, and grasslands are a living community.

China has always been committed to protecting the ecological environment and ecosystems and achieving sustainable development. At present, China has comprehensively established an ecological civilization system consisting of eight systems including the property rights system for natural resource assets and the territorial development and protection system. The *2030 Agenda for Sustainable Development* advocates urgent actions to be taken to address climate change and its impact. In recent years, China has gradually established a system for regulating total consumption and comprehensive conservation of resources, has continued to promote energy conservation and emission reduction in the whole society under the guidance of the Paris Agreement, and has completed the 2020 emission reduction target three years ahead of schedule. China has established the Ministry of Natural Resources as the main body of natural resource assets management to exercise the functions and powers for natural resources protection, development and utilization.

1.1.3 Adhering to New Urbanization as Important as Rural Revitalization

The *2030 Agenda for Sustainable Development* advocates ending poverty and hunger, achieving food security and promoting sustainable agriculture. China has always attached great importance to the issue of agriculture, rural areas and farmers. In recent years, on the basis of promoting new urbanization, China has vig-

Box 1-1 China's Energy Conservation and Emission Reduction Achieved Remarkable Results; 2020 Emission Reduction Target Goal Achieved Three Years Ahead of Schedule

In recent years, China's efforts and achievements in energy conservation and emission reduction and the response to climate change have been remarkable, and have played an important role in promoting global response to climate change. At the 2009 Copenhagen Climate Conference, China proposed that by 2020, the carbon dioxide emissions per unit of GDP would fall by 40% to 45% compared to 2005. By the end of 2017, China had achieved this target goal three years ahead of schedule, and the carbon dioxide emissions per unit of GDP had fallen by 46% compared to 2005. Maria Fernanda Espinosa, president of the 73rd session of UN General Assembly, believes that China's approach is inspiring for many countries and enhances people's confidence in energy conservation and climate change control. According to the data, the proportion of non-fossil energy consumption in China has gradually increased; the total installed capacity of renewable energy power accounts for 30% of the global total; the 2030 target goal for incremental forest reserves has already been achieved ahead of schedule; and electric vehicles in China account for half of the world's total. According to the *Survey Report on Energy Saving and Emission Reduction of Chinese Enterprises since the 13th Five-Year Plan*, nearly 90% of enterprises surveyed can achieve the annual target goals of energy conservation and emission reduction every year. China's economic development is moving away from the stage of high pollution, high carbon emissions, high input and low efficiency and entering a stage of green, low-carbon and high-quality sustainable development.

Volunteers Publically Advocate Energy-saving, Emission-reducing, Low-carbon and Environmentally-Friendly Lifestyle by Riding Bicycles (Photo by Han Suyuan) (Source: http://www.cnsphoto.com)

orously implemented the strategy of rural revitalization, adhered to the priority development of agriculture and rural areas, established and improved the holistic system and policy framework for integrated development of urban and rural areas and accelerated the modernization of agriculture and rural areas in accordance with the general requirements: building rural areas with thriving businesses, pleasant living environments, social etiquette and civility, effective governance, and prosperity. The promulgation of the *Views of the CPC Central Committee and the State Council on Implementing the Rural Vitalization Strategy* and the Strategy Plan for Rural Vitalization (2018-2022) laid out this blueprint and implementation path for rural revitalization.

1.2 Urbanization Level and Quality

China has always been committed to providing all citizens with equitable and inclusive education and medical services, improving infrastructure and improving the quality of the ecological environment to achieve sustainable development. In recent years, China's urbanization level and quality have improved significantly, the household income has increased steadily, and basic public services and infrastructure levels, ecological environment and energy efficiency in urban areas have all continuously improved.

1.2.1 China's Urbanization Rate Reaches Average Level of Middle-income Countries

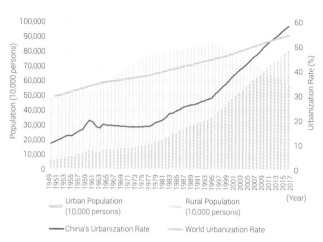

Figure 1-1　Schematic Diagram of a Comparative Analysis of Urbanization Development Levels between China and the World
(Source: China Statistical Yearbook 2018, World Urbanization Prospects: 2018 Revision)

According to United Nations statistics, from 1950 to 2018, the total urban population of the world increased by 3.47 billion. China's urban population has increased from 65 million in 1950 to 837 million in 2018, registering an increase of 770 million, which accounted for more than one-fifth of the world's total.

From 1950 to 2018, the world urbanization rate increased by 0.38 percentage points per year, while China's urbanization rate increased rapidly by an average of 0.7 percentage points per year, about twice the world average. Especially since the reform and opening up launched in 1978, China's urbanization rate has increased by an average of 1 percentage point per year, far exceeding the world average of 0.4 percentage points. In 2011, the national urbanization level of China exceeded 50%. In 2013, China's urbanization level reached 53.73%, surpassing the global average. In 2017, China's urbanization level reached 58.52%, reaching the average level of middle-income countries. At the same time, the living standards of urban residents in China have increased year after year. From 2012 to 2017, the per capita disposable annual income of urban households has increased steadily from RMB 24,565 to RMB 36,396, an increase of about 50%. China is advancing the largest urbanization process in human history, which is profoundly affecting and changing the global urbanization pattern.

1.2.2 Living Quality and Standards of Urban and Rural Residents in China Improves Significantly

In recent years, the social security, medical care and education levels of urban and rural residents have steadily increased. In 2016, the number of people covered by the basic urban medical care insurance reached 744 million, accounting for 53.8% of the permanent resident population. In 2017, China set up the nationwide social security card system, and residents are now entitled to medical and social security services throughout

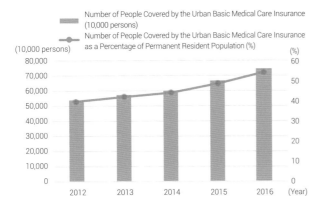

Figure 1-2 Number of People Covered by the Urban Basic Medical Care Insurance in China from 2012 to 2016
(Source: China Statistical Yearbook)

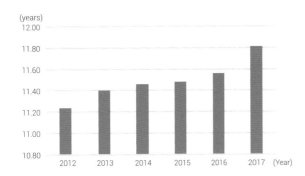

Figure 1-3 Average Years of Schooling for the Working-age Population from 2012 to 2017
(Source: China Statistical Yearbook)

the country by simply presenting the card. The new-type rural cooperative medical care insurance has essentially achieved full coverage and effectively reduced the medical burden on farmers. In 2012-2017, the number of health technicians per thousand residents in the cities nationwide increased from 8.54 persons/1000 residents to 10.87 persons/1000 residents. The average years of education for the working-age population (aged 15-64) continue to increase and was approaching 12 years by 2017, which is equivalent to a high school education. The coverage of affordable housing for urban permanent residents reached 22.6% in 2016, an increase of 10.1 percentage points compared with 2012.

1.2.3 China Adopted Human-centered Transformation Practices to Improve Livability Level of Cities

The *New Urban Agenda* advocates systematically solving urban problems via multi-subject human interactions in the following three types of areas: social, economic and environmental. In recent years, some cities in China have aimed for the goals of livability, harmony and vitality, and have therefore opened up a human-centered transformation method to improve their city's livability levels and development vitality.

Beijing proposed the development goals of building a world-class metropolis that is harmonious and livable. Through the special operation of upgrading through function transfer, the Municipality increased efforts to relieving Beijing of industrial functions nonessential to its role as the capital city. The spaces thus vacated are now being used to supplement living spaces as well as being used to house public services and infrastructure. What's more, this increases the "greening" of more land, and helps to develop high-grade, precision and ad-

Figure 1-4 Xizongbu Hutong, Upgraded Through Adjustments, Where Residents Return to Slow Pace of Life in Quiet Alley (Photo by Yuan Zhou)
(Source: http://www.cnsphoto.com)

Box 1-2 *Shanghai Street Design Guidelines* Promotes Human-centered Transformation of Streets

Streets are an important medium through which people feel the city. In Shanghai, where they have essentially stopped putting up larges numbers of installations, more refined and human-centered street design is an important way to improve the livability of the city. The guidelines propose that the transformation from "road to street" in urban planning and design should be completed in terms of concept, technology and evaluation. By promoting the convergent development of streets and neighborhoods, adopting slow-traffic-first as the principle of street design, and reshaping the urban texture through micro-updating of the communities, we will create a safe and pleasant street space, preserve the history and temperament of a city, and stimulate a city's vitality.

In May 2019, Shanghai Fengyuli Stone Gatehouses Completed Micro-renewal of City Blocks and Renewed its Vitality (Photo by Wang Gang)
(Source: http://www.cnsphoto.com)

vanced industries, thus improving urban environmental quality, and upgrading urban functions.

Shanghai proposed to build a city of happiness and humanity with cultural attractiveness, and took the lead in proposing a 15-minute life circle concept in China, issuing the *Shanghai Planning Guidance of 15-Minute Community-Life Circle* and *Shanghai Street Design Guidelines*. These guides were put into place in order to realize, though the improvement of living units, a transformation of urban lifestyle and the planning implementation of community management in the new era. At present, Shanghai is carrying out pilot projects on planning and construction of life circles in various communities to promote a humanistic transformation of Shanghai streets.

1.3 Overall Pattern of Urbanization

1.3.1 Influence of Chinese Cities in the World Urban System Continues to Rise

In the process of urbanization development in China, with urban agglomerations as the main urbanization pattern, i.e. metropolitan areas as the core carriers, central cities at its core, and coordinated development of large, medium and small cities and small towns, rural revitalization has gradually formed.

China's population and economic activities continue to be concentrated in large cities and urban agglomerations. At the beginning of the 'Reform and Opening Up' period, there were only 2 megacities and 16 large cities in the whole of Mainland of China. By 2018, there were already 2 megacity behemoths, 13 megacities and 105 large cities. China's 19 urban agglomerations account for 20% of China's territorial land area. They have also gathered about 70% of China's urban population and contribute to 76% of the country's GDP.

The status of Chinese cities in the global urban system has continued to rise. According to the global cities ranking published by the Globalization and World Cities (GaWC) Research Network, China's cities have displayed the largest increase in size and number from 2010 to 2018. Hong Kong, Beijing, Shanghai, Taipei, Guangzhou and Shenzhen ranked among the world-class cities in 2018.

1.3.2 Three Major Urban Agglomerations Lead with Regional Cooperative Development Promoted In-depth

The smooth implementation of national-level regional strategies including the Yangtze River Economic Belt, the coordinated development of Beijing-Tianjin-Hebei, the regional integration of the Yangtze River Delta, and the Guangdong-Hong Kong-Macao Greater Bay Area has effectively supported this urbanization development.

The development strategy of the Yangtze River Economic Belt continues to exert its strength. The Yangtze River Economic Belt, which is composed of 11 provinces and cities and accounts for about one-fifth of China's territorial land area, contributes to more than two-fifths of China's total economic output. In 2012-2017, the GDP of the Yangtze River Economic Belt has grown at an average annual rate of 8.6%, and the region has become an important supporting belt for China's overall economic development.

The coordinated development of Beijing, Tianjin and Hebei has been carried out in an in-depth and orderly manner. Within the five-year implementation, the Beijing-Tianjin-Hebei coordinated development

Box 1-3 Planning and Design Concepts of Xiongan New Area

Xiongan New Area is the anchor for "relieving Beijing of functions non-essential to its role" as the capital according to the comprehensive arrangements of CPC Central Committee and the State Council. It is an innovative development demonstration area that implements the new development concepts. Together, along with the second administrative center of Beijing City, Xiongan New Area will form two new wings for Beijing to promote the Beijing-Tianjin-Hebei coordinated development. According to the requirements of the CPC Central Committee

Self-driving Micro-circulation Electric Bus Debuts at Xiongan Citizen Service Center in Hebei
(Photo by Luo Yunfei)
(Source: http://www.cnsphoto.com)

and the *Master Plan for Xiongan New Area in Hebei Province (2018-2035)*, the Xiongan New Area will be built into a green ecological livable new city, an innovation-driven development-leading zone, a demonstration area for coordinated development, and a pioneering zone of open development and shaping of urban style in a Chinese genre, complete with lake sceneries and innovative orientations. According to the plan, Xiongan New Area will undertake transferred functions nonessential to Beijing's role as the capital in an orderly manner, optimize the development pattern of land space, create a beautiful natural ecological environment, and promote integrated development of urban and rural areas. It will also shape the landscape features of the new area, creating a livable environment by constructing a modern integrated transportation system; building a green low-carbon, world-class, and innovative city; and creating a city of digital intelligence to ensure safe operations within the city. The Xiongan Citizen Service Center is the first urban construction project initiated after the establishment of the Xiongan New Area. It is the first project to present the functional orientation and development concepts of the new area and is mainly used to serve the citizens and administrative organs in the provision of temporary office and living places for enterprises moving into the new area in its early stage. At the same time, it will meet such functional needs as becoming a place to display achievements made in the new area in planning and construction, provide government services, offer a space for conference reception and be the center representing future low-carbon smart city life experiences.

Figure 1-5 On May 23, 2019, Visitors Operate Flight Simulator System at first Yangtze River Delta Integration Innovation Achievements Exhibition
(Photo by Zhao Qiang)
(Source: http://www.cnsphoto.com)

Figure 1-6 On October 24, 2018, Hong Kong-Zhuhai-Macao Bridge Officially Opened to Traffic. Buses Pass Near Qingzhou Waterway Bridge
(Photo by Zhang Wei)
(Source: http://www.cnsphoto.com)

strategy promoted the coordinated regional development by relieving Beijing of functions nonessential to its role as the capital, and, after the formation of the new pattern of capital development featured by the symbolism of "one body" (the central area for political function) and its "two wings" (the secondary administrative center and Xiongan New Area of Hebei Province), the "Beijing-Tianjin-Hebei" track has gradually taken shape .

The establishment of the Xiongan New Area in April 2017 in Hebei Province was a major historically strategic choice made by the CPC Central Committee with Comrade Xi Jinping at its core. It is a millennium plan and a national event. In the past two years, the Xiongan New Area has been planned with the guiding features of "high starting point, high standard and high level", aiming at creating an innovative development demonstration area that implements new development concepts. At present, the Xiongan New Area has entered the construction stage.

Since 2016, Beijing has planned to build the second administrative center of Beijing in accordance with a global vision, international standards, Chinese characteristics, and high-profile positioning, which will form the second of the two new wings of Beijing with the first being the Xiongan New Area of Hebei. Beijing's Administrative Center officially moved into Tongzhou District, Beijing on January 10th, 2019.

The regional integration process of urban agglomerations in the Yangtze River Delta has accelerated. In November 2018, the regional integration of the Yangtze River Delta was upgraded to a national strategy. In recent years, various regional cooperation platforms in the Yangtze River Delta region have emerged, covering investment, science and technology, public services and other fields. Furthermore, the institutional systems in the fields of medical care, education and social security are also accelerating. In 2018, the pilot program for non-local direct settlement of outpatient expenditures in the Yangtze River Delta was officially implemented; nine cities in the G60 Science & Technology Innovation Valley in the Yangtze River Delta implemented unified online government services. More convenient public services are provided based on one universal card and the recognition it provides of one governmental seal to meet the needs of the people in

Box 1-4 Beijing Plans High Level Construction of Second Administrative Center

Beijing Urban Master Plan (2016-2035) proposes to advance the planning and construction of Beijing's second administrative center by adhering to a global vision, international standards, Chinese characteristics, a high starting point, the most advanced concepts, the highest standards, and the best quality. This will be accomplished while aiming to create a historical project of artistic value and building a demonstration area for the creation of a world-class harmonious and livable city, as well as a demonstration area for both new urbanization and Beijing-Tianjin-Hebei regional coordinated development. The *Regulatory Plan for the Beijing Second Administrative Center (Blocks) (2016–2035)* proposes that the second administrative center of Beijing should be built into a green city, a forest city, a sponge city, a smart city, a humanistic city, and a livable city, so that the second administrative center will become a new landmark in the capital. The Plan proposes the building of an urban spatial structure featuring "one belt, one axis, and multiple clusters", which highlights the urban features of convergent water and a city landscape, blue waters and green

Citizens Entertain Themselves in Tongzhou Grand Canal Forest Park in Beijing (Photo by Zhu Zhenqiang)
(Source: http://www.cnsphoto.com)

trees, as well as cultural heritage; It also demonstrates a shaping of the capital city genre into one with the charm of the Grand Canal, a humanistic style, and in accordance with modern styles. With the capacity of its environmental resources forming the hard constraint, the capital will maintain strict control over the size of the center, and undertake the functions of the central city in an orderly manner. The second administrative center takes the administrative offices, business services and cultural tourism as its leading role, forming comprehensive urban functions with sound services and building a future urban area free of "urban diseases" while also promoting the coordinated development of the second administrative center and the three counties north of Langfang, Hebei Province.

the pursuit of a better life[①].

Significant progress has been made in the construction of the Guangdong-Hong Kong-Macao Greater Bay Area. The Guangdong-Hong Kong-Macao Greater Bay Area includes nine cities in the Pearl River Delta region of Guangdong and two special administrative regions of Hong Kong and Macao. It is one of the regions with the highest degree of openness and the strongest economic vitality in China. The Mainland has signed investment agreements and economic and technological cooperation agreements with Hong Kong and Macao respectively, marking a new stage of economic and trade exchanges between the Mainland, Hong Kong and Macao. Significant progress has been made in infrastructure construction in the Greater Bay Area. On September 23, 2018, the Hong Kong section of the Guangzhou-Shenzhen-Hong Kong high-speed rail was opened for operation and passengers can now depart from Hong Kong and reach dozens of cities such as Beijing and Shanghai via high-speed rail. On October 24, 2018, the world's longest sea-crossing bridge project, the Hong Kong-Zhuhai-Macao Bridge, was officially opened to traffic, and provides a convenient route for traffic between these three places which were previously separated by the sea.

1.3.3 Metropolitan Areas Become New Entry Point for Improving Quality and Upgrading China's Urbanization

In February 2019, China issued the *Guiding Opinions on Cultivating the Development of Modern Metropolitan Areas*, which stipulates that in order to promote urban integration of central cities and surrounding cities (towns), efforts shall be made to cultivate and develop a number of modern metropolitan areas with innovative institutional mechanisms as the starting point, focusing on promoting unified market development, integrated and highly-efficient infrastructure, the shared development and use of public services, industry specialization and division of labor, joint protection and control of ecological environment, and the convergent development of urban and rural areas. This is so as to form new regional competitive advantages and provide important support for high-quality development of urban agglomerations, economic transformation and upgrading.

---

① He Xinrong, Qu Lingyan, Yang Shaogong, etal. Pursuit of Integration and Restarting - Observation of New Progress in Yangtze River Delta Integration. [EB/OL]. (2019-01-27)[2019-03-06]. http://www.xinhuanet.com/fortune/2019-01/27/c_1124049261.htm.

Box 1-5 Breakthroughs Achieved in Multiple Fields in Development of Nanjing Metropolitan Area

In recent years, the Nanjing metropolitan area has made breakthroughs in such fields as infrastructure, industrial cooperation, ecological environment, and public services. In terms of infrastructure, Nanjing Lukou International Airport has established off-site terminals in the 7 cities of the metropolitan area. The Nanjing-Chuzhou Bus Hub has been put into operation, and the inter-city bus routes from Nanjing to Ma'anshan, Jurong (of Zhenjiang) and Yizheng (of Yangzhou) have been opened to solve the last one-kilometer issue of residents' travel. In terms of industrial cooperation, Nanjing and Yangzhou jointly created the Nanjing-Yangzhou Industrial Belt for Green New Materials and thus strengthened industrial cooperation. In terms of its ecological environment, an environmental event information exchange and consultation mechanism, environmental emergency response and consultation mechanism, environmental temporary control and emergency response mechanism for major activities and events have all been established. They have also realized much regional joint prevention and control and the continuous improvement of the ambient fine particle air quality. In terms of public services, the bus payment system of Nanjing, Zhenjiang, Yangzhou and Ma'anshan has been unified. The coverage of bus cards will be expanded to promote the interconnection of such fields as public transportation, subway, tourism and medical care, and ticketing services for passenger transportation to and from major cities in the metropolitan area has also been networked[①].

Second Session of Nanjing Metropolitan Area Cultural Talents Fair held in Nanjing on March 30, 2019 (Photo by Liu Xiaochu)
(Source: http://www.cnsphoto.com)

---

① Mao Qing. 8 cities including Nanjing and Wuhu to build national-level metropolitan area. [N/OL]. Nanjing Daily, 2018-12-24[2019-03-06]. http://njrb.njdaily.cn/njrb/html/2018-12-24/content_522986.htm?div=-1.

1.4 Approach to Urbanization Development

1.4.1 Development of Secondary and Tertiary Industries Creates More Job Opportunities; Development of Industry-city Integration Greatly Promoted

With the continuous advancement of the industrial and service industries, the number of jobs in China has been increasing, and the scale of employment has continued to expand, which has beneficially supported urbanization development. In 2017, there were 222,000 private industrial enterprises above a designated size, and employed people were numbered at 32.71 million.

The service industry has become the main force in creating job opportunities for residents and the proportion of employment in this tertiary industry has increased from 12.2% in 1978 to 44.9% in 2017. Rapid development of modern information technology such as the Internet+ has improved the efficiency of economic operations and greatly facilitated and enriched both human productivity and lives.

Since 2015, relevant state authorities have successively issued policies and opinions to guide the healthy development, transformation and upgrading of new state-level districts, development zones, and industry-city integration demonstration areas. Under the guidance of national policies, various provinces and municipalities have successively introduced detailed

---

Box 1-6 Kunshan Development Zone's industry-city Integration Achieved Remarkable Results

Covering an area of 115 square kilometers, with an urbanization rate of over 90%, Kunshan Economic & Technological Development Zone is home to more than 2,360 foreign-funded enterprises from 51 countries and regions with a total investment of 39.6 billion US dollars. In recent years, with the great development of the industries, the industry-city integration has promoted the "new leap" of the Kunshan Development Zone. In terms of urban construction, the city promoted urban renewal and reconstruction represented by Qingyang Port waterfront city center, the plots along Chaoyang Road and the urban portal function area of the South Railway Station. In terms of investment in education, the annual expenditure on education exceeds 700 million yuan, and has enabled the building and expansion of schools, introduction of famous teachers and principals, and an improvement of the quality of education. Many migrant workers have settled in this zone, and their children account for 3% of the enrollments of local public schools. In terms of social security, the zone has vigorously raised the level of social security services, improved basic social security systems covering the five insurances of pension, medical care, unemployment, work-related injury, and maternity, and has continued to issue micro-credits for 13 years to help more than 2,000 farmers start their own businesses. These micro-credits added up to a total loan amount of 350 million yuan. In terms of ecological project development, the green area coverage of the developed area is over 42%. 30 km of riverside green corridors, nearly 1,000 acres of sports parks and nearly 800 acres of wetland parks have been built[1].

---

[1] Yin Li Bo. Deciphering the Grand Dvelopment of Kunshan Development Zone, the 5th largest national-level development zone. [EB/OL]. (2019-01-15)[2019-07-18]. http://isuzhou.me/2019/0115/2390912.shtml.

policies and implementation strategies, which have effectively promoted the development of this industry-city integration.

1.4.2 Innovation and Entrepreneurship Promotes Employment Growth; Opening to the Outside World Enhances Development Momentum

The *2030 Agenda for Sustainable Development* advocates the use of technological innovations to increase productivity and advocates for policies that encourage entrepreneurship and stimulate employment. China attaches great importance to innovation, entrepreneurship and employment. Since the initiation of "mass innovation and entrepreneurship", the innovations vitality of the whole society has been stimulated and the scale of employment has been expanded. Since 2016, China has undergone two waves of initiating the construction of 120 "doubling up on creation" or mass innovation and entrepreneurship demonstration bases in three fields including geographic regions, universities and research institutes, and enterprises. In 2018, there were 6.7 million newly registered enterprises nationwide, and 18,000 new enterprises were established every day throughout the year, up by 10.3% and 8.43% respectively on a year-by-year comparative basis. The number of market entities exceeded 100 million, and the number of new jobs in cities and towns nationwide was 13.61 million. The focus of support for innovation and entrepreneurship in various places has shifted to creating innovation resources sharing platforms. Additionally, the 120 mass innovation and entrepreneurship demonstration bases have gradually become regional innovation highlands, which has resulted in more than 6,900 mass innovation spaces and more than 4,800 technology business incubators[①].

After years of strenuous efforts, the gap between China and developed countries in terms of technological innovation capabilities has been narrowing. According to the 2018 Global Innovation Index (GII) report, China's GII ranking has risen from 34 in 2012 to 17 in 2018. The rate of contribution from technological innovations to economic development has steadily increased from 52.2% in 2012 to 57.5% in 2017.

Establishing a fair, open and inclusive global trading system is an important goal advocated by the international community. In line with the new trends of global economic and trade development, China has been more proactive in opening up to the outside world and has contributed to the world by building pilot free trade zones and holding import expos. With the initiation of the region-wide exploration at the Hainan Free Trade Zone in 2018, the number of China's pilot free trade zones reached 12, and Pilot Free Trade Zones

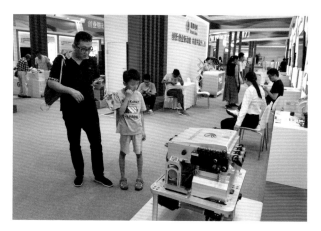

Figure 1-7 Hydrogen Engine Attracts Visitors at 2019 National Mass Innovation and Entrepreneurship Week (Photo by Wang Gang)
(Source: http://www.cnsphoto.com)

---

① National Development and Reform Commission of the People's Republic of China. Officials of Department of High-Tech Industry, NDRC answer questions on 2019 National Mass Innovation and Entrepreneurship Week. [EB/OL].(2019-05-27)[2019-07-09]. http://www.ndrc.gov.cn/gzdt/201905/t20190527_936855.html.

Box 1-7　Shanghai Import Expo Opens the New era of a Shared Future

With the theme of "New Era, Shared Future", the first China International Import Expo is China's practical action towards promoting an open world economy and supporting economic globalization. It has attracted 172 countries, regions and international organizations and more than 3,600 companies to participate in the expo, and more than 400,000 domestic and foreign buyers attended the conference to negotiate purchases. The total exhibition area reached 300,000 square meters and the turnover reached US$57.8 billion. Behind the fruitful results of the import expo, global companies are optimistic about the prospects of China's further expansion and opening, as well as China's market with its more than 1.3 billion population. By now, 36 Fortune-500 companies and leading enterprises of different industries have collectively signed up to participate in the 2nd China International Import Expo in 2019.

At First China International Import Expo, Visitors Visit Large-scale Construction Machinery Exhibitor (Caterpillar) Booth (Photo by Du Yang) (Source: http://www.cnsphoto.com)

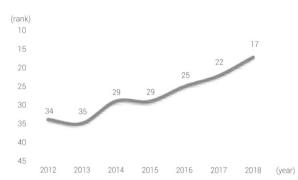

Figure 1-8　China's Global Innovation Index (GII) ranking, 2012-2018
(Source: 2018 Global Innovation Index (GII))

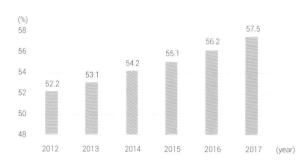

Figure 1-9　Contribution Rate of Technological Innovations to Economic Development in China, 2012-2017
(Source: China Statistical Yearbook on Science and Technology)

have played an important demonstration and guidance role in promoting the trade development between China and world and achieved positive results. The first China International Import Expo was held in Shanghai from November 5th to 10th, 2018. It is the world's first import-themed national exhibition and is an important pioneer in the history of international trade.

1.4.3　Explorations in Green Development Go Deeper; Green Ways of Working and Living Gradually Popularized

Green development is the core concept guiding China's development of an ecological civilization, and has become an important mission in promoting development of new types of urbanization in China.

The concept of green transport and commuting is gradually being promoted. China vigorously supports priority development of public transport and has optimized and improved green transport systems for slow traffic. In 2017, China's urban public transport system saw passenger traffic of 127.34 billion individualized single uses throughout the year, bus lane mileage reached 11,000 kilometers, and the urban bus (trolleybus) lines reached 57,000. In recent years, the central and local governments have actively promoted the development of the safe, convenient and comfortable pedestrian and slow cycling traffic systems in order to encourage green transport or ecological commuting. Many cities such as Xiamen and Beijing have built bicycle lanes to improve the green commuting environment in the region. Bicycle-sharing has also risen rapidly to facilitate citizens' commute in those final kilometers.

A transformation to green ways of working and living is being promoted. With the full launch of various circular economic plans and pilot projects, a total of 49 national-level "urban mining" demonstration bases, 100 pilot demonstration projects of recycling transformation for parks, 100 pilot cities of kitchen waste resource utilization and harmless treatment, 101 national-level circular economy demonstration cities (counties), 28 national-level circular economy demonstration bases of circular economy education have been identified. The central government has successively issued policy documents encouraging green consumption, and has formulated more stringent green consumption standards and management measures[①]. China has vigorously promoted the development of green buildings: by the end of 2016, a total of 7,235 projects nationwide had

---

① National Development and Reform Commission. State New Urbanization Report 2016[M]. Beijing: China Planning Press, 2017.

Figure 1-10 Beijing Opened First bicycle lane, Serving 11,600 Commuters Living Along the Lane (Photo by Jia Tianyong)
(Source: http://www.cnsphoto.com)

obtained green building evaluation labels, with a construction area of more than 800 million square meters.

1.5 Urbanization Supporting Systems Reformed

1.5.1 Full Implementation of Residence Permit System Promotes Settlement of Floating Population in Urban Areas

China has actively promoted the reform of the household registration system, further expanded the channels for permanent household registration, fully lifted or relaxed the restrictions on the permanent household registration of key population groups, and implemented the residence permit system, which has achieved remarkable results. Since 2012, China has completed the registration of more than 80 million members of migrant agricultural population as urban residents. More than 80% of the migrant farmers' children are studying in public schools, and the number of migrant workers receiving vocational skills training has exceeded 20 million.

1.5.2 Reform of Three Types of Lands in Rural Areas Grants Farmers More Property Rights

In recent years, China has steadily promoted the reform of the rural collectively-owned property rights system, launched pilot programs on the reform of the systems of rural land acquisition, enabled the entry of collectively-owned rural commercial construction land and rural housing land into the market, and promoted the separation of rural land ownership, contracting rights, and management rights, thus giving farmers more property rights. Following these new measures, by 2018, 33 pilot counties (cities and districts) had carried out 1275 cases of rural land acquisition, covering an area of 180,000 mu; more than 10,000 plots

Box 1-8   Migrant Farmers' Children: Zero-threshold and Barrier-Free Admission into Schools

8 middle and primary schools in Guta District, Jinzhou City, Liaoning Province officially opened on March 1st. All schools accelerated the adoption of the policy of enrolling migrant farmers' children based on their residence permit, and endeavored to serve migrant workers. Thousands of children of migrant workers have been enrolled in the local schools, and the zero-threshold and barrier-free admission has been highly praised by people from all walks of life.

First day of school, Parents Bring Their Children to Zhongtun Primary School, the Place Where Their Residence Permit is Registered (Photo by Li Tiecheng)
(Source: http://www.cnsphoto.com)

of collectively-owned commercial construction land has been put on the market, covering an area of over 90,000 mu, and involving the total contractual price of about 25.7 billion yuan, the adjustment funds of 2.86 billion yuan, and 228 mortgage loans on collectively-owned commercial construction land were handled with the contractual value of 3.86 billion yuan; isolated and abandoned homesteads involved about 140,000 households and an area of 84,000 mu; and 58,000 cases of rural mortgage loans involved 11.1 billion yuan.

1.5.3   Housing Security System Improves to Meet Housing Needs of Low-and Middle-income Families

The *2030 Agenda for Sustainable Development* advocates ensuring safe and affordable housing for all people and transforming shanty towns. In recent years, China has attached great importance to housing security for low-and middle-income groups, and has successively introduced relevant policies to accelerate the cultivation and development of the housing rental market, further standardizing the development of public rental housing, establishing and improving a long-term mechanism for the stable and healthy development of the real estate market, and accelerating the transformation of shanty towns. In 2018, more than 6.2 million shanty town housing units were renovated nationwide. The public rental housing supply capacity has continued to improve; eligible new employees and migrant workers have been included in the scope of the housing supply, and low-income families with housing problems have been provided with basic public housing.

Chapter 2

Spatial Planning and Urban Governance

The Process of Spatial Planning Reform

Territorial Spatial Planning System

Urban Governance

China Practice: Redevelopment of Underused Urban Land

>> 2

Spatial Planning and Urban Governance

Promoting the integration for urban planning, land use planning and other types of planning and also building a spatial-planning system is an important agenda of reform for the Chinese government in recent years. It's aimed to unify the spatial boundaries of various types of planning and build both an efficient geographic information platform and public management platform in order to serve sustainable development. In March 2018, the first session of the 13th National People's Congress passed the Institutional Reform Plan of the State Council, decided to set up the Ministry of Natural Resources, and demanded to terminate the previous coexistence and conflict pattern of various former planning versions and establish a unified territorial spatial planning system. The new spatial planning system is aimed at building a beautiful China, achieving high-quality development and high-quality life, and promoting governance capacity and the modernization of the national governance system. Future urban planning will continue to exert its governance effectiveness as part of this territorial spatial planning, and likewise, spatial planning will also become an important tool for the spatial governance as well as urban governance.

2.1 The Process of Spatial Planning Reform

2.1.1 Reasons for Reform: Fragmentation of Spatial Governance

Before the institutional reforms of 2018, China had a large and complex planning system. Among other comprehensive plans, the most important included the national economic and social development plan that under the National Development and Reform Commission; the urban and rural planning that under the Ministry of Housing and Urban-Rural Development; and the land use planning under the Ministry of Land and Resources. In addition, there are special plans such as the environmental protection plan under the Ministry of Environmental Protection and the forest land protection and utilization plan under the State Forestry Administration, etc. According to partial statistics, there are more than 80 types of statutory plans. Although their common goal is to guide and limit the behavior of using space, each authority has its own focuses and concerns. For example, the urban and rural planning focuses on urban and rural development management, land use planning focuses on protection of cultivated land, and forest land protection and utilization plan focus on forest land protection. In the process of implementing these plans, various authorities conducted separate decision-making, separate licensing, and separate law enforcement, which led to the fragmentation of spatial governance. The fragmentation of spatial governance leads to two consequences: insufficient protection of natural resources and the lengthy process of administrative decision-making. In 2013, the Third Plenary Session of the 18th CPC Central Committee proposed the reform goals of building the ecological civilization and modernization of national governance system and governance capacity, which makes the spatial planning reform imperative.

2.1.2 Exploration of Reform: the Practice of Multiple Plans Integration

In view of the problems caused by the juxtaposition of multiple planning, "multiple plans integration" (i.e. combining multiple plans into one for unified management) has been proposed as an improvement solution. The early exploration of multiple plans integration began in 2003 with the practice of integrating three plans covering the development, urban, and land use planning piloted by the National Development and Reform Department in Qinzhou, Guangxi and other places. At that stage, the reform was an attempt by a single authority, and its effectiveness and impact were limited. After 2006, the large cities represented by Shanghai and Guangzhou initiated the practice of integrating urban and land use planning based on the merger of local planning with land management authorities and also based on the opportunity for preparation between urban and land use planning. The reform at that stage was the spontaneous explorations of local governments in response to the fragmentation of spatial governance, and showed a certain demonstrative effect. After 2013, under the background of comprehensively deepening a commitment to reforms, spatial planning reform has risen to a national strategy, and the state has promoted a pilot program on cross-departmental integration of multiple planning at the local level. The reform at this stage is a top-down authorized operation. The pilot program mainly includes the multiple planning integration on a pilot basis in 28 cities and counties as well as 9 provincial spatial planning (multi-

Box 2-1 Urban Planning and Land Use Planning

Currently, urban planning and land use planning are the two most important types of legal statutory spatial planning in China. Among them, urban planning is based on the Urban Planning Law (replaced by the *Urban and Rural Planning Law* in 2008), and headed by the Ministry of Housing and Urban-Rural Development (formerly the Ministry of Construction) and is divided into different levels such as urban system planning, overall planning, and detailed planning, etc. It mainly manages the construction activities in the planning area. Land use planning is based on the *Land Administration Law*, headed by the Ministry of Land and Resources, and is divided into five levels: national, provincial, municipal, county, and township (town). It mainly exercises control on the amount of cultivated land exists in the administrative area, the amount of cultivated land occupied for construction, and the total amount of construction land. Both urban master plan and land use master plan must be submitted to the higher-level people's government for approval, and master plan for important cities must be submitted to the State Council for approval.

Land within and without the city (Yichang, Hubei) (Photo by Liu Fengjun)
(Source: http://www.cnsphoto.com)

Prior to institutional reform, land use control within and without the city was divided into two plans and headed by two authorities. In order to achieve alignment and coordination, major cities such as Beijing, Shanghai, and Guangzhou have adopted measures to merge urban planning and land management departments and coordinate urban planning and land use planning.

ple planning integration) pilot projects.

In August 2014, the National Development and Reform Commission, the Ministry of Land and Resources, the Ministry of Environmental Protection, and the Ministry of Housing and Urban-Rural Development jointly issued the *Notice on Conducting the Pilot Work on Multiple Planning Integration in Cities and Counties*, which listed 28 cities and counties across China in the pilot list for spatial planning reform. This was in hopes of promoting multiple planning integration covering economic and social development plan, urban and rural planning, land use planning, ecological environmental protection planning, and creating a merger of these multiple plans into one blueprint of a city (or county).

On the basis of the multiple planning integration pilot program at the city and county level, the *Pilot Program for Spatial Planning at the Provincial Level* was printed and issued in January 2017 by the General Office of the CPC Central Committee and the General Office of the State Council, and the Program included nine provinces in the pilot list for spatial planning reform. It demanded the exploring of and delimiting of urban, agricultural and ecological spaces and their ecological red lines, permanent basic farmland, and urban development boundaries. On this basis, various types of spatial planning are coordinated and to prepare the unified provincial spatial planning. The multiple plans integration pilot program at the city and county level and the provincial-level spatial planning pilot program have reached much consensus regarding the three lines of China's ecological red line agenda, permanent basic farmland, urban development boundaries, and building a unified information platform. It has also provided the foundation for the upcoming full reform.

2.1.3 Direction of Reform: Unified Utilization Control

Although local governments may have different considerations for multiple plans integration, as defined in the central government's agenda for comprehensively deepening the reform, spatial planning reform has always been closely linked to ecological civilization development, natural resources regulation, and the utilization control of territorial space. In 2013, the *Decision of the CPC Central Committee on Several Important Issues of Comprehensively Deepening Reform* first proposed in the chapter titled "Accelerating the Development of the Ecological Civilization System" that: Efforts shall be made to improve the system of natural resource property rights and the system of natural resource utilization control, establish a spatial planning system, and implement the utilization control measures; improve the natural resource oversight system and unify the power over territorial space utilization. In 2015, the *Integrated Reform Plan for Promoting Ecological Progress* established the spatial planning system as one of the eight basic systems for ecological civilization development and required that unified spatial planning be formulated and the multiple plans integration at city and county levels be promoted. In 2018, the *Decision of the Central Committee of the Communist Party of China on Deepening the Reform of the Party and State Institutions* further proposed reforming the natural resources and ecological environment management systems, and required the establishment of state-owned natural resource property management and natural ecological oversight institutions to unify all powers on territorial space utilization control and ecological protection and remediation. It further requires these institutions to strengthen the guidance and binding role of territorial spatial planning for each special plan, advance the multiple planning integration, and achieve organic integration of land use planning and urban-rural planning. It is on the basis of this last document that the Ministry of Natural Resources was established and has assumed the responsibility of "establishing a spatial planning system and supervising its implementation".

In March 2018, the Ministry of Natural Resources was established to integrate the duties of the former Ministry of Land and Resources, the duties of the National Development and Reform Commission for organizing the main functional regions planning, and the duties of the Ministry of Housing and Urban-Rural Development for urban and rural planning and management. Furthermore, it is responsible for unifying and exercising the duties of natural resource asset owners, owned wholly by the people, and all duties of territorial space utilization control and ecological protection and restoration, while also focusing on solving problems such as the absence of natural resources owners and the overlap of spatial planning. It is foreseeable that the spatial planning reform promoted by the Ministry of Natural Resources will focus on the effective supervision of natural resources based on the unified control of territorial space utilization.

2.2 Territorial Spatial Planning System

2.2.1 Overall framework of the territorial spatial planning system

The central government identifies the various types of former spatial planning that have now been unified as territorial spatial planning. In May 2019, the *Opinions of the Central Committee of the Communist Party of China and the State Council on Establishing the Territorial Spatial Planning System and Supervising Its Implementation* was issued, and requested that a territorial spatial planning system and its supervised implementation shall be established. Furthermore, it requested that various types of planning including the main functional regions planning, land use planning, and urban-rural planning, be merged to become unified territorial spatial planning, achieving multiple planning integration and strengthening the guidance and binding role of territorial spatial planning for each special planning mission.

In order to integrate the various types of preexisting planning systems, the new territorial spatial planning reform created a "five-level and three-class" spatial planning system. First of all, territorial spatial planning is being compiled at the five administrative levels of the state, province, city, county, and township; secondly, territorial spatial planning is divided into three categories: overall planning, detailed planning and related special planning. Among them, the state, the province, the city, and the county compile overall spatial planning and the localities compile the township spatial planning in combination with actual situations. The national and provincial level spatial master plans integrate the original territorial planning, land use planning, main functional regions planning, urban system planning, etc.; and the urban and county level spatial master plans primarily integrate the original land utilization planning with both urban and rural planning. The detailed planning is derived from the detailed planning of the city, and also includes village planning. The related special planning refers to special arrangements for the protection and utilization of spatial development in specific areas (watersheds) or specific areas in order to demonstrate special functions, and other special planning involving space utilization. The spatial master plan is the basis for detailed planning and the foundation for relevant special plans.

2.2.2 Objectives of Building the Territorial Spatial Planning System

Based on the functions of the Ministry of Natural Resources, the main objectives of the construction of the territorial spatial planning system include: ① guaranteeing the implementation of "unified exercise and control of all territorial utilization and ecological protection and restoration"; ② promoting the construction of the territorial development and protection system; ③ playing a leading role in the spatial governance system consisting of spatial planning, utilization control, auditing of leading officials on natural resources assets and their depletion, differential performance appraisals, etc.

According to the deployment of the *Opinions of the Central Committee of the CPC and the State Council on Establishing the Territorial Spatial Planning System and Supervising Its Implementation*, by 2020, China will basically establish a territorial spatial planning system by gradually establishing the planning and approval system and implementation supervision system, the regulatory policy system and the technical standard system featured by multiple planning integration. Basically, China will also complete the compilation of the territorial master plans at or above the city and county levels, and initially form a "singular plan" of the territorial development and protection. By 2025, China will strengthen the territorial spatial planning regulations and policies and technical standards system; comprehensively implement the territorial spatial monitoring and early warning and performance appraisal mechanisms; and form a territorial development and protection system with territorial spatial planning as the basis and unified utilization control as the means. By 2035, China will comprehensively upgrade the level of territorial governance system and governance capacity, and basically form a territorial pattern with intensive and efficient production spaces, livable and moderate living spaces, and ecological spaces with lucid waters and lush mountains, security and harmony, strong competitiveness and sustainable development.

2.2.3 Governance Improvement of Territorial Spatial Planning System

Territorial spatial planning will establish a nationally unified, scientific and efficient planning system with clearly defined obligations and powers. In order to improve the planning and governance mechanisms, spatial planning will be improved in the following five aspects:

(1) The planning authority shall be strengthened: Once the plan is approved, no competent authorities or individuals may arbitrarily modify or violate the rules set out in the planning. The lower-level spatial planning should be subject to the higher-level one, and the relevant special planning and detailed planning should be subject to the master plans. The principle of planning before implementation must be maintained, and the multiple plans integration shall be carried out with no other spatial planning compiled beyond the territorial spatial planning system.

(2) The planning approval shall be improved: The reviewing and filing system of the territorial spatial planning shall be established hierarchically and supervised by authorities granting approvals to the planning. Efforts shall be made to streamline the content of the approval of the planning, and the supervisory authorities shall only review and approve application under their jurisdiction, which will significantly reduce approval time. The number of cities needing to report to the State Council for approval shall be reduced, while the territorial master plans of the municipality directly under the Central Government, the municipalities with independent planning status, the provincial capital cities and the designated cities of the State Council shall all be examined and approved by the State Council.

(3) The utilization control system shall be improved: Based on territorial spatial planning, the utilization control shall be implemented on all zones and types of territorial space. Construction within the urban development boundaries shall adopt the control mode of "detailed planning + planning permit"; and construc-

tion outside of urban development boundaries shall be zoned according to dominant use and adopt the control mode of "detailed planning + planning permit" as well as "compulsory indicators + zoning access". A special protection system shall be implemented for protected natural areas, important sea areas and islands, important water sources and cultural relics with national parks as the main body.

(4) The implementation of the plans shall be supervised: A dynamic monitoring, assessment, early warning and implementation supervision mechanism for the territorial spatial planning shall be established and improved while maintaining reliance on the basic information platform for territorial space. The competent authorities of natural resources at the higher levels shall, cooperating with the relevant authorities, organize the supervision and inspection on the implementation of the control requirements for various types of control boundaries. These authorities will also organize compulsory indicators in spatial planning from lower-level authorities, and incorporate the implementation status of spatial planning into the records of natural resources law enforcement inspections. Efforts shall be made to improve the long-term mechanisms for monitoring and early warning regarding natural resources and environmental load capacity, establish a regular assessment system for territorial spatial planning, carry out regular assessment on the results in combination with the status quo and planning of national economic and social development, and maintain dynamic adjustments and improvements to spatial planning.

(5) The reform to delegate power, streamline administration and optimize government services shall be advanced: Based on multiple planning integration, efforts shall be made to coordinate the three major links of planning, construction, and management, and promote multiple review integration and multiple certification integration; optimize the existing approval processes, including the pre-examination of land (or sea) use for construction purposes, site selection, construction land planning permits, construction project planning permits, etc., and improve the efficiency of examination, approval and level of supervision services.

2.3 Urban Governance

2.3.1 Public Participation in Urban Governance

Ensuring satisfactory public participation during the planning phase is conducive to building social consensus and reducing the cost of planning operations and urban governance. The *Urban and Rural Planning Law* of 2008 stipulates to the planning publicity system: "Before submitting an urban or rural plan for examination and approval, the authority in charge of its formulation shall, in accordance with the law, publish the draft of the plan and solicit opinions from experts and the general public by holding appraisal conferences or public hearings, or by other means. Publication of the draft shall remain available for at least 30 days. The authority in charge of plan's formulation shall fully consider the opinions of the experts and general public and, when submitting the materials for examination and approval, attach an explanation and/or reasons for why it adopted relevant opinions." In addition to the planning publicity system, public participation in recent years has gradually extended into the whole process of planning. Since 2014, Beijing and Shanghai have tried open-door planning in the process of compiling the latest version of the urban master plans, that is, ensuring the full and multi-faceted participation of citizens, experts, other

Figure 2-1 Public Representatives of Xiamen Participate in the Environmental Impact Assessment Symposium for Planning (Photo by Yang Fushan)
(Source: http://www.cnsphoto.com)

Figure 2-2 Publicity of Regulatory Planning of Xiong'an Pilot District, Hebei Province (Photo by Han Bing)
(Source: http://www.cnsphoto.com)

governmental authorities, and governments at all levels under the leadership of the planning authorities.

The channels of public participation in the daily urban governance are also expanding. The *Several Opinions of the Central Committee of the CPC and the State Council on Further Strengthening the Management of Urban Planning and Construction* of 2015 required that urban management be transformed into urban governance, and proposed that efforts be made to standardize the scope, rights and channels of public participation in urban governance according to law. It also cleared the channels for public participation in urban governance in an orderly manner; advocated urban management volunteer services; established and improved systems and organizational coordination mechanisms for the publicity and mobilization, organization management and incentive support (etc.) of urban management volunteer services; guided exchanges and cooperation between volunteers and civil organizations, charities and non-profit social organizations; and organized and carried out diverse and normalized volunteer service activities. These guide-lines also advocated supporting and standardizing the development of service-oriented, public welfare and mutual assistance social organizations in accordance with the law. It further urged efforts to adopt such practices as open-to-public days and themed experiential activities to guide social organizations, market intermediaries, citizens, and legal persons to participate in urban governance, and thus to form an urban governance model whose key feature is multi-governance and beneficial interaction.

2.3.2 Grassroots Community Governance

The core of the modernization of within the national governance system and its capacity is to build a limited but effective government, foster an autonomous and self-service society, improve the free but regulated market order, weaken the top-down management of the government, and strengthen the new types of governance consisting of regulation, consultation, and cooperation. With the rapid advancement of urbanization, grassroots units represented by the community have increasingly become the main social space for residents' daily activities. Strengthening community governance has become an inevitable requirement for this aforementioned modernization. For a long time, government organizations have been the main body of community governance in China, and issues such as the low degree of participation of residents and non-government organizations in community governance as well as the presence of relatively weak community organizations. The improvement of grassroots community governance calls for promoting the construction and development of the community away from "the government" as a single subject and towards the multiple subjects of not just government but also community residents and social organizations, and gradually building a modern community governance model featured by government guidance, community autonomy, and participation of people from all walks of life.

To this end, the *Several Opinions of the Central Committee of the CPC and the State Council on Further Strengthening the Management of Urban Planning and Construction* proposes to improve the urban grassroots governance mechanism; further strengthen the leadership and core role of the sub-district and community-level party organizations; promote the development of both community residents' autonomous organizations and community-based social organizations through developing community service-oriented party organiza-

Box 2-2 Beijing Olympic Village Sub-district Innovate Modes of Residents' Participation in Politics and Autonomous Administration

In 2016, the CPC-government-mass Co-governance Project of Beijing Olympic Village Sub-district extended downward to the level the Residents' Councils and set up 17 community councils, which can solve community problems and strengthen homeland consciousness of residents while also enhancing community cohesion. The Sub-district has formed a three-level deliberation platform consisting of the sub-district, the community and the residential quarters. In 2017, the sub-district also proposed the concept of deepening the commitment to "one core and multiple subjects", integration and co-governance, and exploring the working mode featured by the party's leadership, public participation, co-governance, cross-boundary alliances, cohesion and synergy.

Beijing Olympic Village Sub-district holds council meeting (Photo by Ren Haixia)
(Source: http://www.cnsphoto.com)

tions; strengthen the community service functions, and achieve favorable interactions between government and its governance versus the social regulation of residents' self-governing autonomy. The *Guiding Opinions of the Central Committee of the CPC and the State Council on Deepening the Reform of Urban Law Enforcement System and Improving Urban Management Work* further requires that efforts be made to bring into play the role of community in urban governance, establishing a community public affairs access system in accordance with the law, giving full play to the role of community neighborhood committees, and enhancing the functions of community autonomy. Efforts shall also be made to fully engage the role of professional talents such as social workers; cultivate community social organizations; improve community consultation mechanisms; and promote the formulation of community residents' conventions and advance the residents' autonomous management. Furthermore, the building and improvement of community public service facilities to create a convenient and efficient living circle; and providing convenience-for-people and benefit-for-people public services through such means as the establishment of comprehensive community information platforms, the compilation of urban management service illustrated handbooks, and the establishment of mobile service stations.

2.3.3 Smart City Governance

Scientific and technological progress has provided new technical tools for urban governance. The *Guiding Opinions of the Central Committee of the CPC and the State Council on Deepening the Reform of Urban Law Enforcement System and Improving Urban Management Work* of 2015 proposed to improve urban governance by integrating information platforms and building smart cities:

(1) Integrating information platforms: Efforts shall be made to actively promote the digital, refined and smart transformation of urban management and form a digital urban management platform by integrating all cities and counties by the end of 2017. This comprises comprehensively utilizing the modern information technologies such as Internet of Things, cloud computing, and big data; integrating public facilities information and public infrastructure services covering such things as population, transport, energy, and construction, and expanding the functions of digital city management platforms based on urban public information platforms; and accelerating the upgrade of digital urban management to smart cities management by realizing the five-in-one integration of perception, analysis, service, command and supervision. It also includes integrating the urban management related telephone service platforms to create the 12319 urban management service hotline nationwide whose connection can be secured with the alarm call 110; comprehensively adopting various types of monitoring and control measures to strengthen the comprehensive collection and management analysis of urban operation data such as video surveillance, environmental monitoring, traffic operation, water, gas and power supply, flood and waterlogging control, and lifeline support to form a comprehensive urban management database, and focusing on developing the database on urban buildings and structures; and strengthening the collection and integration of total-factor data of urban management such as administrative licensing, administrative sanctions, and social integrity, etc., to improve data standardization, promote the interconnection and open sharing of multi-sector public data resources,

Box 2-3　Beijing Uses Big Data to Improve Urban Governance

In 2016, the first grassroots government-based big data center of China was set up in the West Chang'an Avenue Sub-district in Xicheng District, Beijing, which removed the barriers between various data platforms and integrated the scattered data on the E-Government Network. Data integration has activated the previously dormant data, and enabled them to exert more precise governance benefits in urban governance and people's livelihood security. The big data platform not only improved security monitoring of the downtown area and surrounding areas, but also optimized the "last metre" for serving citizens.

West Chang'an Avenue Sub-district Big Data Center, Xicheng District, Beijing
(Photo by Han Haidan)
(Source: http://www.cnsphoto.com)

Box 2-4　Haizhou District, Lianyungang City: Creating a Smart City Management Platform

Haizhou District Smart City Management Center of Lianyungang City, Jiangsu Province focuses on the goals of information-based transformation, refined management, humanized services, efficient operation, and integrating the government service hotline with digital urban management, comprehensive governance, people's livelihood services and emergency response, etc. into the smart city management platform based on a comprehensive urban grid management. This not only broadens the information collection channels and ensures participation in community public services in an all-round way, but also promotes urban development based on collaboration, participation, and common interests, while also effectively improving the level of refined urban management.

Haizhou District Smart City Management Center, Lianyungang (Photo by Geng Yuhe)
(Source: http://www.cnsphoto.com)

and establish the new mechanism featured by reasoning based on data, decision-making based on data, management based on data and innovation on data.

(2) Creating smart cities: Efforts shall be made to strengthen the intelligent management and monitoring services of urban infrastructure, accelerate the intelligent transformation and upgrading of municipal public facilities, build an urban virtual simulation system, and strengthen the construction of key urban application projects. Further efforts are needed to develop smart water services, build intelligent water supply and drainage and sewage treatment systems that cover the whole process of water supply and ensure the quality and safety of water supply; develop the smart pipe network to realize the information-based management and intelligent operation of urban underground space, underground utility tunnel and underground pipe network; and to develop intelligent buildings that allows us to realize energy-saving and safe intelligent control of building facilities and equipment. Furthermore, smart cities require the acceleration of the information-based transformation of urban management and comprehensive law enforcement archives; reliance on information technology to comprehensively utilize video integration technology in exploring new law enforcement modes such as rapid disposal and off-site law enforcement, and the improvement of law enforcement effectiveness.

2.4 China Practice: Redevelopment of Underused Urban Land

2.4.1 Concept of Underused Urban Land Redevelopment

One of the purposes of the central government's establishment of the spatial planning system and the harmonization of the territorial utilization is to control the expansion of urban construction land. In the face of land demand brought about by economic development, the central government has always attached importance to the efficient use of stock construction land while strengthening growth control. In as early as 2004, the State Council required local governments to use land economically, strive to revitalize land stocks, and grant the powers to revitalize land stocks and resulting benefits to local governments. In 2008, the State Council reiterated that local governments should pursue economical and intensive use of land, give priority to the development and utilization of vacant, abandoned, idle and underused land, and strive to improve the efficiency of construction land use. In 2009, Guangdong Province promoted the transformation of old towns, old factories, and old villages by taking advantage of the deployment to build a demonstration province for the economical and intensive utilization of land jointly carried out alongside of the Ministry of Land and Resources. The transformation of old towns, old factories, and old villages was later called one kind of "urban renewal" in cities such as Guangzhou and Shenzhen, and "redevelopment of underused urban land" at the national level. The "urban renewal" is similar to the "redevelopment of underused urban land", and both focus on the renovation of old towns, old industrial areas, urban villages, and shanty towns, etc.

In 2014, the Ministry of Land and Resources issued the *Guiding Opinions on Promoting the Economical and Intensive Utilization of Land*. This document required that efforts be made to promote urban renewal and urban land redevelopment in a standardized and orderly manner to enhance the population and industry load-carrying capacity of urban land under the precondition of strictly protecting historical and cultural heritage, traditional architecture and maintaining distinctive features. It also pushed to establish a reasonable interest distribution mechanism in conjunction with the transformation of urban shanty towns, and to adopt such methods and negotiation-based recovery and purchase for reserve, etc., to promote the transformation of "old towns"; handle relevant procedures in accordance with the law, encouraging the transformation and industrial upgrading of "old factories"; and to fully respect the wishes of right holders. This, in turn, encourages the adoption of such methodologies as independent development, joint development, acquisition & development and other means to promote the transformation of urban villages.

In November 2016, the Ministry of Land and Resources issued the *Guiding Opinions on Deepening the Redevelopment of Underused Urban Land (Trial)*, which clarified that the underused urban land refers to stock urban construction land featured by a scattered layout, extensive utilization, unreasonable purposes, and dilapidated buildings, including land for prohibited or outdated industries as specified in the national industrial policies, or land that does not meet the requirements for safe production and environmental protection. It also includes industrial land for implementation of the "phase out the secondary industry to make way for the tertiary industry" strategy; and old urban areas, urban villages, shanty towns, old industrial areas, etc. which have scattered layout and backward facilities, and are facing planned renovations. The goal of redevelopment is to promote urban renewal and industrial transformation and upgrading, and to optimize the land utilization structure, enhance the population and industrial load-carrying capacity of urban construction land, and to build harmonious and liveable towns.

2.4.2 Effectiveness of underused urban land redevelopment

According to the statistics of the Ministry of Natural Resources, as of the end of 2017, seven provinces (or cities) including Shanghai, Jiangsu, Zhejiang, Hubei, Liaoning, Shaanxi, and Guangdong had all started the redevelopment of underused urban land. These 7 provinces (or cities) have identified a total of 413,300 hectares of underused urban land, and completed 148,000 renovation and redevelopment projects within an area of 46,100 hectares, accounting for about 11% of the total identified land areas. The completed projects are mainly concentrated in the three provinces of Guangdong, Jiangsu and Zhejiang which started relevant work in the early period. Among them, Guangdong completed the renovation area of 21,200 hectares, accounting for 8% of the province's identified area of underused land; Zhejiang completed the renovation area of 17,400 hectares, accounting for 23% of the province's identified underused land area; and Jiangsu completed the renovation area of 4,800 hectares, accounting for 23% of the province's identified underused land area. The land renovated is mainly industrial land and residential land. Among the projects completed in various places, industrial land accounts for 51% and residential land accounts for 23% in Guangdong; industrial land accounts for 53% and residential land accounts for 16.7% in Zhejiang; and industrial land accounts for 74.9%, and residential land accounts for 14.3% in Jiangsu.

The effects of underused urban land redevelopment are the following:

(1) The living environment of residents has been improved: Among the completed renovation projects in Guangdong Province, 1201 urban infrastructure and

public welfare projects were built, and 597.35 hectares of public green space were added; 7,770,600 square meters of traditional cultural and historical buildings were protected and restored; and a total of 45,500 sets of affordable housing units were built. The shanty towns of Niushan in Ouhai District, Wenzhou City, Zhejiang Province have been transformed into beautiful new homes with mountains and waters, beautiful environments and complete facilities from the previously dirty, noisy and disorderly residential compounds they once were.

(2) The urban land utilization structure has been optimized: In some industrial plant redevelopment projects, Shanghai has increased its green space percentage to 30% by reducing industrial land and increasing public open space. Liaoyang City of Liaoning Province has used the land from discontinued enterprises to implement renovation and development projects under the "phase out the secondary industry to make way for the tertiary industry" strategy, and increased its public service functions.

(3) The economic restructuring has been promoted: Among the renovated projects in Guangdong Province, there are 3,246 industrial structure adjustment projects, accounting for 60% of the total number of renovation projects, of which 496 fall under the category of outdated and outward transferred projects, and 442 modern service industry and high-tech industry projects have been introduced. Through promoting redevelopment, Zhejiang Province has promoted the elimination of more than 9,000 outdated-production-capacity enterprises, rectified and eliminated more than 85,000 low-end enterprises, and disposed of 882 "zombie enterprises".

(4) The land utilization efficiency has been improved: Among the plots that have been redeveloped

Figure 2-3 Achievements of the Transformation of Zhijiang Cultural and Creative Park in Xihu District, Hangzhou
(Source: Department of Natural Resources of Zhejiang Province)

in Zhejiang Province, the average plot ratio of industrial land has increased from 0.78 to 1.71; investment intensity has increased from 67,000 yuan/hectare to 171,000 yuan/hectare; the output has increased from 57,000 yuan/ha to 183,000 yuan/hectare; and the tax has increased from 8,000 yuan/hectare to 30,000 yuan/hectare.

2.4.3 Patterns of underused urban land redevelopment

In practice, multiple redevelopment patterns have been formed in various places, including via land purchase, reserve and renovation by the government, the self-initiated renovation by the original state-owned land use rights holders, the self-initiated development by the original collective economic organizations, the renovation and development by newly introduced market entities, the joint development by government and social organizations, and other diversified models such as cooperative development by multiple social organizations.

Among them, Guangdong Province mainly adopted the pattern of self-initiated renovation by the original state-owned land use rights holders, which accounts for 34% of the total projects; the pattern of renovation and development by newly introduced market entities accounts for 28%; the pattern of self-initiated development by the original collective economic organizations accounts for 22%; the pattern of land purchase, reserve and renovation by the government accounts for 10%; and the patterns of joint development by government and social organizations and cooperative development by multiple social organizations account for 6%.

Zhejiang Province mainly adopted the pattern of renovation and development by newly introduced market entities, which accounts for 39%; the pattern of land purchase, reserve and renovation by the government accounts for 31%; the pattern of self-initiated renovation and development by the original state-owned land use rights holders accounts for 23%; the pattern of self-initiated development by the original collective economic organizations accounts for 5%; and the pattern of joint development by government and social organizations and cooperative development by multiple social organizations account for 2%.

Jiangsu Province mainly adopted the pattern of land purchase, reserve and renovation by the government, which accounts for 52%; the pattern of self-initiated renovation and development by the original state-owned land use rights holders and the pattern of renovation and development by newly introduced market entities each account for 23%; and the pattern

Box 2-5 Modes of Underused Urban Land Redevelopment

(1) Land purchase, reserve and renovation by the government
The government implements land expropriation, compensation and resettlement, and necessary infrastructure construction through the land reserve institution. After commercial land has become "net land" (or surplus land), it will be assigned through public bidding, auction, and listed for bidding, and the transferee will develop and start construction according to the planning.
(2) Self-initiated renovation by the original state-owned land use rights holders
This pattern is mainly adopted for state-owned industrial land, where original state-owned land use rights holders use such means as conversion of state assets into state shares, lease and joint development, etc. to adjust the industrial structure, convert land utilization purposes, and improve land economic benefits.
(3) Self-initiated development by the original collective economic organizations
The rural collective economic organizations, as the renovation entity, renovate the urban villages through self-raised funds or in cooperation with newly-introduced cooperation units. After the commercial land has become a net land, it can be assigned by listing for pre-determined utilization purposes. Rural collective economic organizations may apply for the conversion of all their land into state-owned construction land.
(4) Renovation and development by newly introduced market entities
Market entities will purchase the land before moving to re-development. Guangdong Province allows market entities to acquire adjacent land parcels, and to merge the land parcels into cadastral parcels for centralized renovation.

of self-initiated development by the original collective economic organizations, the pattern of joint development by government and social organizations and the pattern of cooperative development by multiple social organizations each account for 1%.

Hubei Province mainly adopted the pattern of land purchase, reserve and renovation by the government, which accounts for 70%; the pattern of self-initiated development by the original collective economic organizations accounts for 22%; the pattern of renovation and development by newly introduced market entities accounts for 7%; and the pattern of self-initiated renovation and development by the original state-owned land use rights holders accounts for 1%.

In order to fully mobilize the enthusiasm of the original land rights holders, various measures have been taken to share value-added benefits. Zhejiang grants certain rewards to the original land rights holders through the means of of statutory compensation in case the original land rights holders withdraw or purchase stock construction land for redevelopment. Shanghai allows the original land rights holders to develop the land on their own as a single entity or through joint development by such means as paying land utilization right assignment fees for stock housing launched to the market, and provides that a certain proportion of value-added income shall be paid to original land rights holders for the industrial land that is publicly transferred after having been purchased and put into reserve. Quanzhou of Fujian Province allows the lease of the factory from the original land use rights holders to re-lease the land after the renovation, thus achieving the separation between the land use rights and the land management rights and benefit sharing.

Chapter 3

Urban Infrastructure

Relevant National Plans and Policies

Urban Transport System

Urban Water System

Urban Energy System

Urban Sanitation System

Urban Communication System

China Practice: Smart City

>> 3

Urban Infrastructure

Urban infrastructure is the material basis for new urbanization, the basic guarantee for urban social and economic development, and also the basic guarantee for the improvement of human settlements, public services and safe operation of cities. Finally, urban infrastructure is the skeleton and lifeline of urban development. In order to promote the development of urban infrastructure, China persists in planning and guidance, and thus compiled relevant plans including the *13th Five-Year Plan for National Urban Municipal Infrastructure Development*, which lays out plans for infrastructure development. This directive promotes high-quality development of transport facilities, water facilities, energy facilities and environmental sanitation facilities through a series of measures to better meet the needs of the people in the pursuit of a better life and, therefore, supports the development of new urbanization. The rapid development of such technologies as new communication technologies, Internet of Things, big data, cloud computing (among others), has promoted the construction and management of smart cities, and further improved the level of urban operational management and security.

3.1 Relevant National Plans and Policies

3.1.1 Comprehensive Plan for Municipal Infrastructure

This is the first time that China has prepared such a national-level comprehensive municipal infrastructure construction plan, the *13th Five-Year Plan for National Urban Municipal Infrastructure Development*, which was released by the National Development and Reform Commission and the Ministry of Housing and Urban-Rural Development in May 2017. This plan covers the following seven areas: urban transport system, underground pipeline system, water system, energy system, environmental sanitation system, green space system and smart city. It also defines the development goals and construction tasks of the national urban infrastructure during the *13th Five-Year Plan* period. The plan proposes that by 2020, a modern urban municipal infrastructure system with reasonable layout, complete facilities and functions, and high security and efficiency will be built to meet the needs of a moderately prosperous society. The capacity of infrastructure to support economic and social development will be greatly enhanced. The main development indicators of urban municipal infrastructure during the *13th Five-Year Plan* period include: the establishment of an interconnected road traffic network, the goal for the road network density in urban built-up areas to reach 8 km/km^2; the further expansion of the scope of public water supply services as well as the penetration rate of public water supply in cities officially designated in China to reach over 95%; the expansion of the application field and scale of natural gas, as well as the gas penetration rate in cities officially designated in China to reach over 97%; the

Figure 3-1 The first underground utility tunnel in Yichang, Hubei Province
(Photo by Zhang Guorong)
(Source: http://www.cnsphoto.com)

significant improvement of urban water environmental quality, as well as the control of the black and odorous water in built-up areas of prefecture-level and above cities to be below 10%; the assurance that the coverage of parks and green spaces within the accessible radius of residential areas be no less than 80%; the promotion of utility tunnel construction in an orderly manner and the comprehensive allocation rate of utility tunnels for national urban roads to reach approximately 2%, as well as the establishment of a batch of underground utility tunnels meeting advanced international standards with reasonable layouts, well-arranged entry lines, highly efficient operations, and orderly management to be put into operation; the construction of sponge cities to be accelerated, and 20% of urban built-up areas to meet construction requirements of sponge cities.

3.1.2 Plan for Sewage Treatment and Recycling Facilities

The *13th Five-Year Plan for the Construction of National Urban Sewage Treatment and Recycling Fa-

cilities, released by the National Development and Reform Commission and the Ministry of Housing and Urban-Rural Development in December 2016, clarifies the seven tasks of China's sewage treatment and recycling facilities during the 13th Five-Year Plan period, including improving the sewage collection system, improving the capacity of sewage treatment facilities, attaching importance to non-hazardous treatment and disposal of sludge, promoting the use of reclaimed water, initiating the pollution control of initial rainwater, strengthening the comprehensive rectification of urban black and odorous water bodies, and strengthening the building of regulatory capacity, etc. At the same time, the plan puts forward corresponding construction goals for each work content: during the *13th Five-Year Plan* period, China will add 125,900 kilometers of sewage pipe network, renovate 27,700 kilometers of old sewage pipe network, renovate 28,800 kilometers of combined pipe network, increase the processing capacity of sewage treatment facilities by 50.22 million m^3/day, upgrade and retrofit the processing capacity of sewage treatment facilities by 42.2 million m^3/day, increase the harmless disposal capacity of sludge (wet sludge of 80% moisture content) by 60,100 tons/day, increase the processing capacity of wastewater recycling facilities by 15.05 million m^3/day, and increase the capacity of initial rainwater treatment facilities by 8.31 million m^3/day. Efforts shall also be made to strengthen supervision capacity building and initially form a national unified urban drainage and sewage treatment supervision system with full coverage.

Figure 3-2　Sihong of Jiangsu Province uses the tail water of the sewage treatment plant to build a wetland into an ecological park (Photo by Zhang Lianhua) (Source: http://www.cnsphoto.com)

3.1.3 Plan for Environmental Sanitation Facilities

The *13th Five-Year Plan for the Construction of Non-hazardous Treatment Facilities for Municipal Solid Waste*, released by the National Development and Reform Commission and the Ministry of Housing and Urban-Rural Development in December 2016, defines six tasks for China's construction of environmental sanitation facilities during the *13th Five-Year Plan* period, including accelerating the construction of treatment facilities, improving the garbage collection and transport system, enhancing the management of stocks, promoting the resource utilization and non-hazardous treatment of food waste, promoting the sorting of municipal solid waste (MSW), and strengthening regulatory capacity building. During the *13th Five-Year Plan* period, the construction goals include: the increase of the incremental non-hazardous treatment capacity of MSW by 509,700 tons/day (including the capacity of 129,000 tons/day from the continued construction brought forward by the *12th Five-Year Plan* period); the increase of the MSW incineration capacity as a percentage of total non-hazardous treatment capacity in cities officially designated in China to reach 50%, and 60% for the Eastern Region; to increase the incremental garbage collection and transport capacity to 442,200 tons/day; to implement 803 stock management projects; to up the incremental food waste processing capacity to 34,400 tons/day; the fundamental establishment of food waste reclamation and recycling systems in cities; the enhancement of regulatory capacity building; and the initial formation of a relatively complete MSW treatment regulatory system.

3.1.4 Planning for Transport Facilities

Issued by the State Council in February 2017, the *13th Five-Year Plan for the Development of Modern Comprehensive Transport System* provides an integrated development plan in such areas as infrastructure layout, strategic support, transport services integration, development of intelligence transport, green transport development, emergency response and guarantee system, etc., and proposes multiple construction tasks including multi-connected comprehensive transport links and high-quality rapid transport networks, etc. By 2020, a modern, comprehensive transport system that is safe, convenient, efficient, and green will be built and some regions and areas will take the lead in realizing transport modernization. The major indicators for the development of a comprehensive transport network during the *13th Five-Year Plan* period include: the coverage of high-speed rails to be more than 80% of cities with permanent urban resident population of over 1 million; the basic coverage of railways, expressways, and civil aviation transport airports in cities with a permanent urban resident population of over 200,000; the doubling of the operating mileage of urban rail transits compared with figures from 2015; the increase of the total mileage of the comprehensive transport network to reach around 5.4 million kilometers; and the closer connection of methods of transport modes to reach 1-2 hours of commuting time between core cities as well as between those core cities and neighboring node cities in important urban agglomerations.

3.1.5 Plan for Communication Facilities

In December 2016, the Ministry of Industry and Information Technology released the *Information and Communication Industry Development Plan (2016-2020)* and the *Information and Communication Industry Development Plan - Internet of Things (2016-2020)*, which proposes the following development goals by 2020: to fully realize the objectives of the "Broadband China" strategy; to build the basic infrastructure for the new generation of high-speed, mobile, secure and ubiquitous information traffic; to form the initial networked, intelligent, service-oriented and synergistic modern Internet industrial system; to essentially form the initial internationally competitive Internet of Things industrial system; and to ensure the overall scale of industries including perception manufacturing, network transmission, and intelligent information services will exceed 1.5 trillion yuan.

3.1.6 Renovation of Old Communities

The old urban residential areas are the "last mile" of municipal infrastructure for cities and directly relate to the peoples' daily life, but they are often the weak link of infrastructure. In April 2019, the Ministry of Housing and Urban-Rural Development, along with the National Development and Reform Commission and the Ministry of Finance, invited 31 provinces, autonomous regions and municipalities, as well as Xinjiang Production and Construction Corps to carry out comprehensive investigations. As of the end of May, the reported number of old urban residential areas that needed to be renovated was 170,000, which involves hundreds of millions of residents. By the end of 2017, the Ministry of Housing and Urban-Rural Development launched pilot projects for the renovation of old urban residential areas in 15 cities including Xiamen and Guangzhou. As of December 2018, the pilot cities had transformed 106 old urban residential areas, benefiting 59,000 households and producing replicable ways to advocate renovations. In 2018, more than 10,000 elevators had been installed in old urban residential areas across China, more than 4,000

Figure 3-3 3 sets of elevators added to existing residential buildings in Sanxinjiayuan, Hangzhou, Zhejiang Province passed the acceptance test
(Photo by Shi Jianxue)
(Source: http://www.cnsphoto.com)

are under construction, and more than 7,000 are in the early stages of installation. Since 2019, the Ministry of Housing and Urban-Rural Development, along with the National Development and Reform Commission and the Ministry of Finance, have carefully studied the supporting policies for the renovation of old urban residential areas, and issued the *Notice on Doing a Good Job in Renovating the Old Urban Residential Areas in 2019* to comprehensively stimulate the transformation of old urban residential areas.

3.2 Urban Transport System

3.2.1 Regional Transport

(1) Planning and Construction of China High-speed Railway (HSR) Network

In 2018, the operating mileage of China's high-speed railways was 29,000 kilometers, accounting for more than two-thirds of the world's total mileage of high-speed railways. It is planned that by 2020, on the basis of the completed "four vertical and four horizon-

Box 3-1 Key High-speed Railway Projects for the 13th Five-Year Period[①]

Beijing-Shenyang HSR, Beijing-Zhangjiakou-Hohhot HSR, Datong-Zhangjiakou HSR, Shijiazhuang-Jinan HSR, Jinan-Qingdao HSR, Zhengzhou-Xuzhou HSR, Baoji-Lanzhou HSR, Xi'an-Chengdu HSR, Shangqiu-Hefei-Hangzhou HSR, Wuhan-Shiyan, and Nanchang-Zhangzhou HSR shall be completed.

Shenyang-Dunhua, Baotou-Yinchuan HSR, Yinchuan-Xi'an HSR, Beijing-Shangqiu HSR, Taiyuan-Jiaozuo HSR, Zhengzhou-Jinan HSR, Zhengzhou-Wanzhou HSR, Huanggang-Huangmei HSR, Shiyan-Xi'an HSR, Hefei-Anqing-Jiujiang HSR, Xuzhou-Lianyungang HSR, Chongqing-Qianjiang HSR, Chongqing-Kunming HSR, Guiyang-Nanning HSR, Changsha-Ganzhou HSR, Ganzhou-Shenzhen HSR, and Fuzhou-Xiamen HSR shall be constructed.

tal" grid network of high-speed railways[②], the construction of the "eight vertical and eight horizontal" passageway grid of high-speed railways[③] will be launched, and 30,000 kilometers of high-speed railways will be built, covering more than 80% of large cities. The travelling time between Beijing and most provincial capital cities shall be kept between 2 to 8 hours, travelling time between neighboring large and medium-sized cities shall be maintained between 1-4 hours, and the 0.5-2-hour convenient commuting within major urban agglomerations shall be realized.

From 2017 to 2018, the Baoji–Lanzhou HSR, Xi'an-Chengdu HSR, Shijiazhuang-Jinan HSR, Hong Kong section of Guangzhou-Shenzhen-Hong Kong HSR, Hangzhou-Huangshan HSR, Harbin-Mudanjiang HSR, Jinan-Qingdao HSR, Huaihua-Hengyang, Jiujiang-Quzhou HSR, and Kunming-Dali HSR had started operation. On September 21, 2017, the Fuxing bullet trains took the lead in running at a speed of 350 km/h on the Beijing-Shanghai high-speed railway. China has become the country with the highest speed of commercial operation of high-speed railways in the world.

---

① Extract from the *13th Five-Year Plan for Railway Development*, issued by the National Development and Reform Commission (NDRC) with other ministries or central governmental organs.
② Four vertical tracks: Beijing-Shanghai HSR, Beijing-Hong Kong HSR, Beijing-Harbin HSR and Hangzhou-Shenzhen HSR. Four horizontal tracks: Shanghai-Wuhan-Chengdu HSR, Xuzhou-Lanzhou HSR, Shanghai-Kunming HSR and Qingdao-Taiyuan HSR.
③ Eight vertical passageways: Coastal Passageway, Beijing-Shanghai Passageway, Beijing-Hong Kong (Taiwan) Passageway, Beijing-Harbin ~Beijing-Hong Kong- Macao Passageway, Hohhot-Nanning Passageway, Beijing-Kunming Passageway, Baotou-(Yinchuan)-Haikou Passageway and Lanzhou (Xining)–Guangzhou passageway. Eight horizontal channels: Suifenhe–Manzhouli Passageway, Beijing–Lanzhou Passageway, Qingdao–Yinchuan Passageway, Eurasia Continental Bridge Passageway, Yangtze River Passageway, Shanghai–Kunming Passageway, Xiamen–Chongqing Passageway and Guangzhou–Kunming Passageway.

Figure 3-4 On July 11, the direct high-speed trains between Chongqing and Hong Kong officially started operation (Photo by Chen Chao)
(Source: http://www.cnsphoto.com)

(2) Construction of transport facilities in urban agglomerations of Beijing-Tianjin-Hebei region, Yangtze River Delta and Greater Bay Area.

In 2018, Beijing-Xiong'an Intercity Railway and Xiong'an Station started construction. According to the plan, Xiong'an Station will be connected to five railway lines including Beijing-Shangqiu High-speed Railway, Beijing-Xiong'an Intercity Railway and Tianjin-Xiong'an Intercity Railway and is expected to be completed and opened to traffic in 2020. The construction of Beijing Daxing International Airport, Yanchong Expressway and Beijing-Xiong'an Expressway has been accelerated. The Beijing-Tianjin-Hebei regional comprehensive transport network with high-speed railways and intercity railways as its skeleton is rapidly coming into being, supporting the coordinated development of Beijing-Tianjin-Hebei region and the development of Xiong'an New Area. On June 30, 2019, Daxing International Airport was completed.

In 2018, the railway construction of the Yangtze River Delta was rapidly advanced: high-speed railways including Shangqiu-Hefei-Hangzhou HSR, Lianyungang-Huai'an-Yangzhou-Zhenjiang HSR, Xuzhou-Suqian-Huai'an-Yancheng HSR, Lianyungang-Xuzhou HSR and the coastal high-speed railway along the Yangtze River in south Jiangsu were under construction and the construction plan of inter-city railways for the coastal urban agglomerations in Jiangsu Province was approved[1]. By the end of 2018, the total mileage of high-speed railways in Yangtze River Delta had reached 4,171 kilometers, connecting 34 cities above the prefecture level in Shanghai, Jiangsu, Zhejiang and Anhui Provinces.

Remarkable progress has been made in the transport network development of Guangdong-Hong Kong-Macao Greater Bay Area. The 55-kilometer long Hong Kong-Zhuhai-Macao Bridge, a combination of bridges, tunnels and islands, was completed and opened to traffic on October 24, 2018 and is a major land transport project that connects Mainland China, Hong Kong and Macao. Chinese President Xi Jinping attended the opening ceremony and announced the official opening of the bridge. The 24-kilometer Shenzhen-Zhongshan Bridge spanning the Pearl River Estuary is also under construction and is expected to be completed and opened to traffic by 2024. By that time, the second large-scale passageway connecting the two sides of the Pearl River Estuary and promoting the coordinated industrial development of Guangdong, Hong Kong and Macao will also be formed.

3.2.2 Public Transport

(1) Trolley and bus[2]

In 2017, there were 651,200 trolleys and buses in China, with the total length of operating routes

Figure 3-5 Aerial view of Daxing International Airport (Photo by Sun Zifa) (Source: http://www.cnsphoto.com)

---

[1] Total mileage: 1063km, among which the section in Jiangsu Province: 980km, and the section in Anhui Province: 83km.
[2] Source: National Report On Urban Passenger Transport Development 2017, compiled by the Ministry of Transport of the People's Republic of China.

Box 3-2 Development and Changes in Urban Public Transport within 40 years

In the past 40 years of reform and opening up, great achievements have been made in China's urban public transportation development. In 2017, compared with 1978, the number of trolleys and buses increased from 25,500 to 651,200, and there was an average annual growth rate of 8.9%; the length of the operating lines increased from 47,400 kilometers to 1,069,400 kilometers, with an average annual growth rate of 8.55%; and the passenger traffic increased from 12.84 billion single-trip users to 72.287 billion single-trip users, registering an average annual growth rate of 4.65%.

Before 1978, Beijing had only completed the urban rail transit called Subway Line 1, with an operating mileage of 30.4 kilometers. In 1982, Beijing completed Subway Line 2, Shanghai completed Metro Line 1 in 1995, and Guangzhou completed Metro Line 1 in 1999. By the year 2000, there were 4 urban rail transit lines in 3 cities in China, with a total operating mileage of 105.7 kilometers. In the late 1990s, China began to pay attention to urban rail transit and speed up construction. By 2006, 10 cities had started the operation of 21 urban rail transit lines, and the operating mileage increased to 584.1 kilometers. In the past 10 years, China's urban rail transit network has developed and increased rapidly: there are 3.4 times the former number of operating cities, 7.86 times the former number of metro lines, and 8.62 times the former length of operating lines.

reaching 1,069,400 kilometers and a total passenger traffic of 72.287 billion single-trip users. 32 cities have opened BRT lines, which involves 8,802 operating vehicles and the total length of operating routes has reached 3424.5 kilometers. The number of new energy vehicles has been increasing, with the number of operating vehicles now reaching 257,200, and their proportion in total operating vehicles increasing from 27.0% in 2016 to 39.5%.

(2) Urban Rail Transit[1]

In 2017, 165 urban rail transit lines were built and started operation in 34 cities in China, with a total mileage of 5,032.7 kilometers, a total of 3,234 stations and a total passenger volume of 18.481 billion single-trip users. The average passenger traffic per day in Beijing, Shanghai, Guangzhou and Shenzhen are the top four in China. At the same time, the demand for urban rail transit and the scale of projects under construction in China is still huge: There are 254 urban rail transit lines under construction in 56 cities, with a length of 6,246.3 kilometers and a total of 4,150 stations.

Figure 3-6 The first phase of Lanzhou Metro Line 1 was officially opened for trial operation on June 23, 2019 (Photo by Yang Yanmin)
(Source: http://www.cnsphoto.com)

---

[1] China Intelligent Transportation Industry Development Yearbook (2017), compiled by China Intelligent Transportation Systems Association.

3.2.3 Shared Mobility

(1) Bike-sharing

In 2017, the number of registered bike-sharing users increased from more than 18 million in 2016 to more than 200 million. Although bike-sharing facilitates public mobility, the blind expansion, non-standard business models and lack of industry supervision of shared bikes has led to the prominent phenomena of parking confusion , inadequate bike operation maintenance, difficulty in refunding user deposits, and increased urban management costs. Many cities have also begun to restrict bike-sharing access or set the "thresholds" in terms of scale. The failure of such companies as ofo to return user deposits in time has become the public focus. In order to standardize the development of bike-sharing, relevant state ministries and commissions have issued guiding opinions①, and clarified development direction, as well as the topic of management responsibility, rights and interests at the policy level.

Figure 3-7　Bike-sharing for easy travel (Photo by Yan Daming)
(Source: http://www.cnsphoto.com)

---

① The Ministry of Transport and other ministries/organs issued the *Guiding Opinions on Encouraging and Regulating the Development of Internet Bike Rental* in 2017.

(2) Taxi

In 2017, the number of China's conventional "cruising" taxis (referred to as "taxis") was 1,395,800 units, a decrease of 8,200 units compared to 2016. Due to the policy restrictions regarding the development scale of taxis, the advent of ride-hailing (referred to as "net-to-car") has shown a rapid growth trend. The convenience and comfort of ride-hailing have also been favored by users, and the number of users exceeds 200 million. The two criminal cases involving passengers being killed in 2018 threw Didi Chuxing, China's popular ride-sharing service provider, into the spotlight and prompted the state regulatory authorities to enter eight ride-hailing companies in order to carry out special security inspections and urge rectification. Travel safety has always been a core issue in the development of ride-hailing technology. If the certification and management model for ride-hailing drivers is not fundamentally changed, and the scale of ride-hailing mechanisms is allowed to expand in a disorderly manner, secure operation and security supervision of ride-hailing still face great risks.

3.2.4　New Energy Transportation

(1) New Energy Vehicles

China's production and sales of new energy vehicles is ranked top in the world. The government has accelerated the development of pure electric vehicles by introducing relevant policies on infrastructure, safety management, technology research and development, and activating the special license plate for new energy vehicles. In 2017, the sales volume of new energy vehicles reached 777,000, marking an annual increase of 53.3%. Among them, the sales volume of pure electric passenger vehicles was 652,000, an annual increase of 59.6%, accounting for 83.4% of new energy vehicle sales[①]. However, the pure electric vehicle's issue of its unstable cruising ability, low cruising range, imperfect layout of charging facilities, and long waiting time for charging are prominent, which has resulted in worsening the consumer's experience. In 2018, more than 40 cases of spontaneous combustion of electric vehicles occurred in China, which also increased consumers' concerns about the safety of pure electric vehicles.

(2) Charging Facilities

The demand for charging facilities is strong, but the problems of difficult construction and low operational efficiency are common. In 2017, China's electric vehicle-charging pile ratio was only 3.5:1, and the utilization rate of public charging facilities was less than 15% due to fewer deployments and charging piles and long queues. In 2018, the National Development and Reform Commission and other ministries and commissions formulated the *Action Plan for Enhancing the Charging Capacity of New Energy Vehicles*, striving to fully optimize the layout of charging facilities, improve the product quality of charging facilities, and solve the problem of construction of charging piles in old residential areas as well as public charging facilities in urban centers within three years to provide a more efficient and convenient charging service for new energy vehicle users.

3.3 Urban Water System

During the *13th Five-Year Plan* period, the Chinese government will adhere to the problem- and goal-oriented approach, further strengthen the construction of urban water systems, focus on building a multi-level barrier for water supply security, comprehensively rectify urban black and odorous water bodies, strengthen the whole process control of water pollution, establish drainage and local flooding prevention and controlled engineering systems, accelerate the construction of sponge cities, transform the development concepts of the urban water systems, improve the management level, and continuously ensure the secure and stable operation of the urban water systems.

3.3.1 Water Supply Security

By the end of 2017, the national urban comprehensive productive capacity of water supply was 304.750 million m^3/d, the annual total volume of water supply was 59.38 billion m^3, and the number of residents with access to tap water was 483 million. The urban public water supply service scope has been continuously expanded, and the public water supply penetration rate has continued to increase to 95.04%. Compared with the end of the *12th Five-Year Plan* period, the productive capacity of national urban water plants increased by 7.967 million m^3/d and the length

Figure 3-8　Mianyang, Sichuan Province issued the first new energy license plate (Photo by Chen Dongdong)
(Source: http://www.cnsphoto.com)

---

① China Intelligent Transportation Industry Development Yearbook (2017), compiled by China Intelligent Transportation Systems Association.

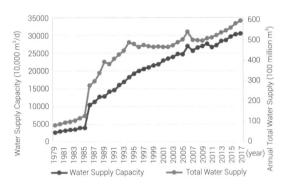

Figure 3-9 Changes in Water Supply Capacity and Total Water Supply in China (1979-2017)
(Source: Author's self-painting)

Figure 3-10 Tourists drink water at the free drinking water fountains in the scenic spot of Guxu Men Ancient City Gate in Suzhou (Photo by Wang Jiankang)
(Source: http://www.cnsphoto.com)

of incremental pipe network reached 87,000 km.

In October 2017, the Ministry of Housing and Urban-Rural Development issued the *Guideline for the District Metered Area Management of Urban Water Supply Distribution System—Construction of Leakage Control System for Water Supply Distribution System (Trial)*, and proposed to take the district metered area management of the urban water supply distribution system as their starting point, strengthening the leakage management and control of the urban water supply distribution systems with systematic ways of thinking, improve the level of refined and information-based management of the distribution system, improve the water supply security support capability, and accelerate leakage control of the urban water supply.

The new version of *Water Pollution Prevention and Control Law* was officially implemented on January 1, 2018. In order to ensure the safety of drinking water for urban and rural residents, the revised *Water Pollution Prevention and Control Law* has explicitly added the "Safety of Drinking Water" regulations into the legislative purpose, and has specifically added a chapter on the "Protection of Drinking Water Sources and Other Special Water Bodies" to further improve the management system of drinking water source protection areas.

By the end of 2018, the national water quality monitoring network of the urban water supply already has one national water quality monitoring center, 43 national stations, and nearly 200 local stations, covering 30 provinces, autonomous regions and municipalities directly under the Central Government. Therefore, the national urban water supply and drainage detection system with its "two-level network, three-level station" system at its core is increasingly being improved upon.

In 2018, the Ministry of Housing and Urban-Rural Development organized and implemented the water supply standardization inspections and water quality inspections. Among them, water quality inspections

Figure 3-11 Leakage inspection in the tap water supply lines (Photo by Niu Zhiyong)
(Source: http://www.cnsphoto.com)

from 2016 to 2018 covered more than 560 cities officially designated in 31 provinces, autonomous regions and municipalities directly under the Central Government, of which 21 province-level administrative regions are fully covered and 10 provincial-level administrative regions partially covered, accounting for over 85% of the total number of cities officially designated in China.

At the National Science and Technology Week held in May 2019, the demonstration model of the full-process drinking water security guarantee system was exhibited in major special exhibition areas and was entitled "Water Pollution Control and Governance." This was organized by the Ministry of Housing and Urban-Rural Development and is the theme of the *National Science and Technology Major Special Drinking Water Security Guarantee*, which demonstrated the latest scientific and technological achievements in the field of drinking water security protection in China in recent years.

3.3.2 Sewage collection and treatment

By the end of 2017, the national sewage treatment capacity reached 157 million m^3/d, and the annual sewage treatment capacity was 46.55 billion m^3. The construction of urban drainage pipelines has obviously accelerated. In 2017, the length of drainage pipelines reached 630,000 kilometers, an increase of 17% over 2015. The reclaimed water treatment capacity reached 35.879 million m^3/d, and the reclaimed water utilization was 7.13 billion m^3.

In April 2019, the Ministry of Housing and Urban-Rural Development, the Ministry of Ecology and Environment, and the National Development and Reform Commission jointly issued the *Three-Year Action Plan for Raising the Quality and Returns of Urban Sewage Treatment (2019-2021)* (MHURD〔2019〕No.52), proposing that after three years of hard work,

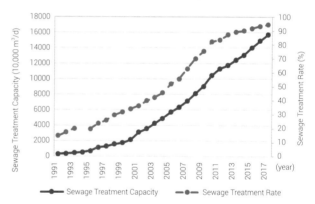

Figure 3-13　Changes in Urban Sewage Treatment Capacity and Sewage Treatment Rate in China (1991-2017)
(Source: Author's self-painting)

Figure 3-12　The demonstration model of the full-process drinking water security guarantee system at the 2019 National Science and Technology Week
(Source: Photo by Author)

Figure 3-14　Fuzhou started black and odorous water body treatment in urban areas (Photo by Zhang Bin)
(Source: http://www.cnsphoto.com)

the built-up areas of prefecture-level and above cities have essentially eliminated indirect discharge of domestic sewage; have fundamentally eliminated empty space domestic sewage collection and treatment facilities in urban villages, old urban areas and urban-rural integration areas; and have, in essence, eliminated the concentration of black and odorous water bodies. The centralized collection efficiency of urban domestic sewage has significantly improved.

3.3.3 Drainage and Local Flooding Prevention and Control

By the end of 2017, the urban rainwater pipe network was 253,600 kilometers long, and the combined rainwater and sewage pipe network was 111,100 kilometers long in 658 cities nationwide. Compared with the end of the *12th Five-Year Plan* period, the length of the urban rainwater pipe network increased by 48,100 kilometers, and the length of the combined rainwater and sewage pipe network increased by 3,400 kilometers. Both have played an important role in ensuring urban drainage safety.

In March 2018 and March 2019, the Ministry of Housing and Urban-Rural Development successively issued the *Notice on Promulgating the List of Responsible Persons for the Security of National Urban Drainage and Local Flooding Prevention and Control and Important Local Flooding Points in 2018* (Letter No.40〔2018〕of the Ministry of Housing and Urban-Rural Development) and the *Notice on Promulgating the List of Responsible Persons for the Security of National Urban Drainage and Local Flooding Prevention and Control and Important Local Flooding Points in 2019* (Letter No.37〔2019〕of the Ministry of Housing and Urban-Rural Development) Document, which proposed to establish an accountability system for the management of important local flooding points in urban areas, and assign the corresponding responsibilities to specific posts and staff.

In March 2019, the Ministry of Housing and Urban-Rural Development issued the *Notice on Doing a Good Job of Urban Drainage and Local Flooding Prevention and Control in 2019* (Letter No.176〔2019〕of the Ministry of Housing and Urban-Rural Development), proposing to plan projects on a dynamic basis, strictly control the quality of the projects, and steadily address the weak links in the process of urban drainage and local flooding prevention and control.

In order to ensure the security of urban drainage and local flooding prevention and control, the Ministry of Housing and Urban-Rural Development and other ministries and commissions have issued relevant requirements in terms of organizational mechanisms, engineering construction, operation and maintenance management, etc., and formulated safeguard measures. However, due to factors such as climate change and historical weakness of urban drainage facilities, the drainage and local flooding prevention and control situation in China is still grim.

Figure 3-15　A worker is maintaining underground pipelines in Anhui
(Photo by Han Suyuan)
(Source: http://www.cnsphoto.com)

3.4 Urban Energy System

3.4.1 Heating Supply

As of 2017, urban central heating area was 8.31 billion m^2, an increase of 920 million m^2 compared with 2016. The length of the heating supply pipelines was 276,300 km, an increase of 62,718 km compared with 2016. The CPC Central Committee and the State Council has attached great importance to clean heating and expressly required multiple times that clean heating should be regarded as an important political task, a task for the livelihood of the people, and an urgent task. The levels of clean heating will be dramatically enhanced to improve the living environment and quality of life for the people.

On September 6, 2017, the Ministry of Housing and Urban-Rural Development and the National Development and Reform Commission jointly issued the *Guiding Opinions on Promoting Clean Heating Supply in Urban Heating Areas in Northern China*, requiring local governments to prepare special plans on heating to accelerate the cleaner use of coal-fired heat sources. On December 5, 2017, the ten ministries and commissions including the National Development and Reform Commission and the National Energy Administration jointly issued the *Clean Winter Heating Supply Plan for Northern China (2017-2021)*, requesting that by 2021, the main urban areas of "2+26" cities in Northern China shall achieve clean heating supply. On December 6, 2017, the National Development and Reform Commission and the National Energy Administration jointly issued the *Notice on Promoting the Development of Biomass Energy Heating* (No.2123〔2017〕of National Development and Reform Commission and National Energy Administration), emphasizing that efforts shall be made to steadily develop MSW incineration cogeneration, accelerate the heating supply renovation of conventional biomass power generation projects, promote the conversion of small thermal power stations to biomass cogeneration, build regional comprehensive clean energy systems, and accelerate the progress of biomass cogeneration technology, etc.; On December 27, 2017, the National Energy Administration issued the *Notice on Doing a Good Job of Clean Heating Supply in the Heating Season of 2017-2018* (Notice No.116 〔2017〕of General Affairs Department of the National

Figure 3-16 An employee of a social welfare center in Rongcheng City, Shandong Province, checks the indoor temperature in the elderly citizen's room (Photo by Yang Meng)
(Source: http://www.cnsphoto.com)

Figure 3-17 Geothermal heating supply is adopted for the shantytown renovation of Sinopec Zhongyuan Oilfield, Puyang City, Henan Province (Photo by Tong Jiang)
(Source: http://www.cnsphoto.com)

Box 3-3 Development of Urban Central Heating Supply Since Reform and Opening Up

Since the reform and opening up, the urban central heating supply capacity nationwide has been significantly improved, and a heat supply pattern dominated by cogeneration of electricity and heat, and supplemented by boiler-heated houses in a few regions has been gradually formed. Among them, the steam heating supply capacity has experienced three stages of development: from 1981 to 1985, the steam heating supply capacity grew steadily and slowly; from 1985 to 1995, the steam heating supply capacity increased rapidly; and from 1996 to 2017, the steam heating supply capacity grew steadily and slowly. The hot water heating supply capacity has experienced two stages of development: from 1981 to 1985, the hot water heating supply capacity grew steadily and slowly; and from 1985 to 2017, the hot water heating supply capacity increased rapidly.

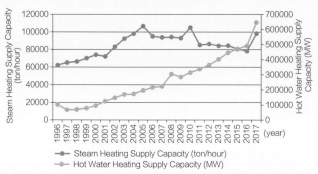

Changes in the Development of Urban Central Heating Supply Capacity (1996-2017)
(Source: Author's self-painting)

Energy Administration), requesting that all local governments shall carry out unified deployment for a clean heating supply and choose diverse and affordable clean heating models according to local conditions.

3.4.2 Gas

The urban gas supply includes manufactured gas, natural gas and liquefied petroleum gas (LPG). With the enhanced awareness of energy conservation and emission reduction and environmental protection in the whole of society, clean and high-heat natural gas energy has received increased attention. By the end of 2017, the total population with access to urban gas in China was 470 million, and the penetration rate of the gas was 96.26%. The total urban manufactured gas supply in China was 2.71 billion m^3, a decrease of 38.6% compared with 2016. The total urban natural gas supply was 126.38 billion m^3, marking an increase of 7.9% from 2016; the total urban LPG supply in China was 9.988 million tons, down by 7.4% from 2016. In general, the supply capacity of urban manufactured gas and LPG has dropped significantly, whereas the natural gas supply capacity has increased significantly, and the proportion of urban natural gas in the urban energy supply has increased.

Figure 3-18 Emergency Response Drill for Leakage of Underground Gas Pipeline Network in Anqing, Anhui Province (Photo by Huang You'an)
(Source: http://www.cnsphoto.com)

Box 3-4 Development of Urban Gas Supply Since the Reform and Opening Up

Overall speaking, China's urban gas supply capacity and gas penetration rate have each increased significantly since the reform and opening up. Among them, the manufactured gas supply capacity has experienced three stages of development: from 1978 to 1989, the manufactured gas supply capacity grew steadily and slowly; from 1989 to 2009, the manufactured gas supply capacity increased

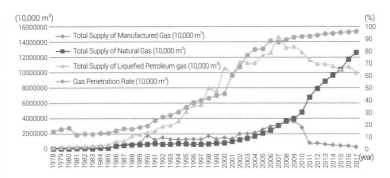

Development and Changes of Urban Gas Supply in China Since the Reform and Opening Up
(Source: Author's self-painting)

rapidly; and from 2009 to 2017, the manufactured gas supply capacity continued to decrease. The natural gas supply capacity has experienced two stages of development: from 1978 to 2000, the natural gas supply capacity grew steadily and slowly; and from 2000 to 2017, the natural gas supply capacity increased rapidly. The liquefied petroleum gas supply capacity has experienced three stages of development: from 1978 to 1985, the supply capacity of liquefied petroleum gas grew steadily and slowly; from 1985 to 2007, the supply capacity of liquefied petroleum gas increased rapidly; and from 2007 to 2017, the supply capacity of liquefied petroleum gas continued to decrease. The national urban gas penetration rate has experienced three growth stages: from 1978 to 1988, the gas penetration rate increased steadily and slowly; from 1988 to 2009, the gas penetration rate increased rapidly; and from 2009 to 2017, the gas penetration rate increased steadily and slowly.

3.5 Urban Sanitation System

3.5.1 Municipal Solid Waste (MSW) Treatment Facilities

(1) The Non-hazardous MSW Treatment Facilities Developing Rapidly

From 2006 to 2017, the nationwide non-hazardous treatment facilities for MSW developed rapidly, and the non-hazardous treatment capacity was significantly improved. The number of non-hazardous treatment stations (plants) increased from 419 in 2006 to 1,013 in 2017, an increase of 1.4 times, with an average annual

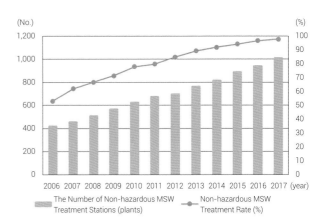

Figure 3-19 Changes of National Non-hazardous MSW Treatment Stations (Plants) and Treatment Rate
(Source: Author's self-painting)

growth rate of 8.4%; and the non-hazardous treatment rate increased from 53.1% in 2006 to 97.7% in 2017, an increase of 0.8 times, with the average annual growth rate of 5.8%.

(2) The Huge Geographical Differences in Development of Non-hazardous Treatment Facilities for MSW

From a national perspective, the average MSW collection capacity nationwide in 2017 was 590,000 tons/day, and the average non-hazardous MSW treatment capacity was 680,000 tons/day. From the perspective of various provinces, Beijing, Jilin, Henan, Chongqing, Sichuan, Tibet, and Qinghai are faced with insufficient capacity for the non-hazardous treatment of MSW.

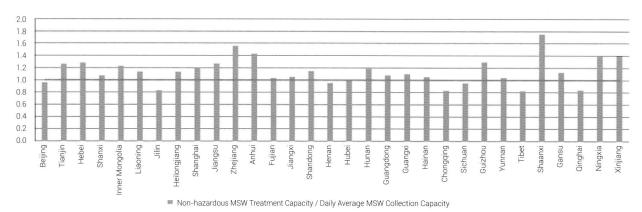

Figure 3-20   Comparison of Treatment Capacity and Daily Average Collection Capacity of National Non-hazardous MSW Treatment Facilities (Source: Author's self-painting)

Box 3-5   Development of Non-hazardous Treatment Facilities for MSW since Reform and Opening up

In general, since the reform and opening up, China's MSW treatment facilities have witnessed rapid development, with the number and capacity of non-hazardous treatment facilities increasing significantly, and the non-hazardous treatment rate continuously improving. Among them, the construction of non-hazardous treatment facilities showed three stages of development: from 1979 to 1989, the number and capacity of non-hazardous treatment facilities developed slowly; from 1989 to 2005, the number and capacity of non-hazardous treatment facilities developed rapidly; and from 2006 to 2017, the number and capacity of

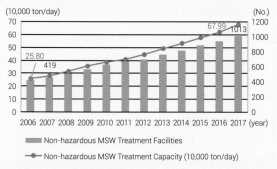

Changes in the Development of Urban Environmental Sanitation Facilities Nationwide Since the Reform and Opening Up (2006-2017) (Source: Author's self-painting)

non-hazardous treatment facilities increased steadily. By the end of 2017, compared with the MSW collection capacity, the non-hazardous treatment rate of MSW is at a relatively high level. As the urban living environment continuously improves, considerable ecological progress has been made in urban areas.

3.5.2 MSW Sorting Pilot Project

In December 2016, the 14th meeting of the Central Leading Group on Financial and Economic Affairs chaired by CPC General Secretary Xi Jinping proposed that "efforts shall be made to accelerate the establishment of the municipal solid waste treatment system characterized by classified delivery, classified collection, classified transport and classified disposal, and includes the building of an MSW classification system that is based on law and promoted by the government, involving people from all walks of life, and is featured by urban and rural coordination but suitable for local conditions, and one that endeavors to expand the coverage of MSW classification". The implementation of MSW classification can effectively alleviate the pressures of transport and terminal treatment, and is of great significance for improving the urban and rural environment, promoting resource recycling, improving the quality of new urbanization, and improving the level of ecological progress.

In March 2017, the National Development and Reform Commission and the Ministry of Housing and Urban-Rural Development issued the *Implementation Plan for the MSW Classification System* (hereinafter referred to as the "Plan"), which laid out a roadmap for the implementation of the MSW classification system.

As of June 2018, more than a year after the implementation of the Plan, MSW classification has been gradually initiated from individual units to whole regions. Public institutions such as the CPC, governmental, and military agencies took the lead in promoting MSW classification. Among them, the classification of MSW carried out by 134 central Party and State institutions has all passed the acceptance tests, and 11 demonstration units have been established. 27 Beijing-based troops have carried out the classification of MSW. Provincial organs of various provinces have generally promoted the classification of MSW, and 29 provinces (autonomous regions and municipalities directly under the central government) have completed the goals of mandatory classification of MSW in provincial organizations.

At the same time, some regions took the lead in establishing a mandatory classification system for MSW. 21 provinces (autonomous regions and municipalities directly under the central government) including the three ecological progress pilot zones of Fujian, Guizhou and Jiangxi have introduced implementation plans for MSW classification. All municipalities directly under the central government, provincial capital cities, cities with separate plans, and some prefecture-level cities are striving to promote the construction of a system comprising of classified delivery, collection, transportation and treatment facilities for MSW. Among them, 41 cities are promoting the construction of MSW classification and accompanying demonstration zones, and 14 cities have introduced

Figure 3-21 Pupils in Hefei City, Anhui Province, learn waste classification
(Photo by Zhang Yazi)
(Source: http://www.cnsphoto.com)

Figure 3-22 A child throws garbage under the guidance of a volunteer at Shanghai Hongqiao Railway Station (Photo by Yin Liqin)
(Source: http://www.cnsphoto.com)

local rules or regulations for the classification of MSW. The sorting of MSW in Xiamen, Shenzhen, Ningbo, Suzhou, Hangzhou, Guangzhou, Shanghai and other cities has achieved initial results. All residential areas within Xiamen Island and 68% of the residential areas with more than 1,000 households on the outskirts of Xiamen Island have adopted MSW classification.

On July 1, 2019, the *Regulations on the Management of MSW in Shanghai* (hereinafter referred to as the "Regulations") was officially implemented, marking the entry of Shanghai into the era of mandatory MSW classification. The *Regulations* divides MSW into four categories: recyclable waste, hazardous waste, household food waste and residual waste. It proposes to implement a total volume control system for the disposal of MSW in Shanghai, and gradually implement the system for the classified dumping of different types of MSW at appointed times and places. The *Regulations* also define the legal responsibilities of individuals, enterprises, MSW collection and transportation organizations, and MSW disposal organizations.

3.6 Urban Communication System

3.6.1 Overall Situation

By the end of 2017, there were 6.19 million mobile phone base stations nationwide, and the length of optical cable lines was 37.8 million kilometers. The penetration rate of mobile phones has continued to rise, and reached 102 units/100 persons by the end of 2017, an increase of 58.4% compared with 2010. The urban land-line telephone penetration rate has continued to decline and reached 18 units/100 persons at the end of 2017, down 42.0% compared with 2010. There are 1.27 billion mobile Internet users nationwide and 350 million Internet broadband users. The plan to boost broadband speeds and lower rates for internet services has been steadily advanced, and breakthroughs have been made in technologies such as 5G and quantum information.

3.6.2 Boosting broadband speeds and lowering rates for internet services

The three major telecom operators, i.e. China Telecom, China Unicom and China Mobile fully canceled domestic long-distance and roaming charges for mobile phones from September 1, 2017.

During the NPC & CPPCC Sessions in 2018, in the government work report Premier Li Keqiang proposed that China should do more to speed up broadband and bring down internet rates, achieve high-speed broadband access in both urban and rural areas, and make free internet access available in more public places. We should significantly lower the rates of home broadband, corporate broadband and dedicated internet access services; domestic data roaming charges should be abolished, and rates for mobile internet services should be cut by at least 30 percent.

During the NPC & CPPCC Sessions in 2019, in the government work report Premier Li Keqiang further proposed that average broadband service rates for small and medium enterprises would be lowered by another 15 percent, and average rates for mobile internet services would be further cut by more than 20 percent. Cell phone subscribers nationwide would be able to keep their numbers while switching carriers, and cell phone packages would be regulated to achieve solid fee cuts for all consumers.

3.6.3 5G Technology

On August 13, 2017, the State Council issued the *Guiding Opinions on Further Expanding and Upgrading Information Consumption to Constantly Release Domestic Demand Potentials*, proposing to accelerate the research, technology testing and industrial promotion of the fifth generation mobile communication (5G) standards, strive to launch commercial operation of 5G in 2020, and efforts shall be made to accelerate the deployment of IOT infrastructure.

In November 2017, the National Development and Reform Commission issued the *Notice on Organizing the Implementation of the Next Generation Information Infrastructure Construction Project in 2018*, proposing to focus on the municipalities directly under the central government, the provincial capital cities, and the major cities in the Pearl River Delta, the Yangtze River Delta and the Beijing-Tianjin-Hebei region to start the construction of 5G network.

In May 2018, the Ministry of Industry and Information Technology and the State-owned Assets Supervision and Administration Commission jointly issued the *Implementation Opinions on Deepening the 2018 Special Action for Speeding up the Broadband and Lowering Internet Rates and Accelerating the Cultivation of New Growth Drivers*, and proposed to promote the standardization, R&D, application, industrial chain maturity and security support of 5G technology, organize the implementation of the major project of new generation broadband wireless mobile communication network, complete the third phase of technology research and development experiments, and promote the formation of a globally unified 5G standard.

According to the 43rd *Statistical Report on Internet Development in China* released by China Internet Network Information Center in February 2019, as of March 2018, China's 5G international standard documents accounted for 32% of the world's total, and the standardization projects guided by such documents accounted for 40% of the world total with the superior speed and driving quality ranking top in the world.

In December 2018, the Ministry of Industry and Information Technology distributed 5G spectrum resources to the three major telecom operators. In June 2019, the Ministry of Industry and Information Technol-

Figure 3-23 Xiong'an unmanned micro-circulation electric bus carries passengers to travel (Photo by Zhang Yuan)
(Source: http://www.cnsphoto.com)

ogy officially issued 5G licenses, making 2019 the first year of China's 5G commercial usage and ushering China into the official era of the fifth generation of mobile telecommunication. At this point, the first 5G phones have now been dialed in all provinces across China.

3.7 China Practice: Smart City

3.7.1 National Smart City Pilot Project

In order to explore the scientific ways of building, operating, managing, serving and developing smart cities, the Ministry of Housing and Urban-Rural Development issued the *Notice on the Pilot Work of National Smart Cities* on November 22, 2012, and launched the pilot project of national smart cities. At the same time, it issued the *Interim Administrative Measures for the Pilot Project of National Smart Cities* and the *National Smart City (District, Town) Pilot Index System (Trial)*, which provided regulations on the reporting, review, creation process management and acceptance of smart cities across China.

Since 2012, the lists of three batches of national pilot smart cities have been published. Among them, in 2012, the Ministry of Housing and Urban-Rural Development determined 90 cities (districts and towns) such as Dongcheng District of Beijing as the first batch of pilot cities; in 2013, the Ministry of Housing and Urban-Rural Development identified 103 cities (districts and counties) such as the Beijing Economic-Technological Development Area / town) as the newly-added pilot cities and other 9 cities (districts, counties, towns) including Xinbei District of Changzhou City as expanded pilot cities; and in 2014, the Ministry of Housing and Urban-Rural Development and the Ministry of Science and Technology jointly organized the national smart city pilot project, added the application channel for special pilot projects, and finally determined 84 cities (districts, counties and towns) such as Mentougou District in Beijing as newly-added pilot cities, 13 cities (districts and counties) such as Zhengding County of Shijiazhuang City, Hebei Province as expanded pilot cities, and 41 projects undertaken by Space Star Technology Co., Ltd. and other companies as national special pilot smart cities.

In August 2014, the eight ministries and commissions including the National Development and Reform Commission issued the *Guiding Opinions on Promoting the Healthy Development of Smart Cities*, proposing that the urban people's government should establish a standardized investment and financing mechanism, and guide social funds to participate in the development of smart cities through various forms such as franchising and purchase of services encouraging eligible enterprises to issue corporate bonds to raise funds for the development of smart cities.

3.7.2 Local Practice Cases

In 2005, Hangzhou City began to implement the exploration and practice of digital urban management. With the overall goals of "fastest speed, widest coverage, optimal functions" and "leading level in China", Hangzhou City strived to promote the development of service-, innovation-and efficiency-based smart city management. In May 2012, the People's Government of Zhejiang Province issued the *Guiding Opinions on Pragmatically Advancing the Demonstration Pilot Project of Smart City Development*, officially listed the development of "Smart City Management" system in Hangzhou as one of the first batch of 13 demonstration pilot projects launched, and opened the prelude to the comprehensive promotion of smart city management in Hangzhou.

The Smart City Management system in Hangzhou takes "one center and four platforms" as the main content, and "coverage and volume expansion, and quality and efficiency improvement" as the main approach to playing an active role in daily management, emergency management, services for the people and scientific decision-making. For more than a decade, Hangzhou's Digital City Management system has registered a total of 15.34769 million urban management cases reported (as of the end of August 2018), and solved 15.3049 million cases, increasing the problem-solving rate from the initial 26.7% to the current 98.87%. Since its launch online in April 2014 (as of the end of August 2018), the APP titled Tiexin Chengguan (urban management by your side) has released information concerning 424 urban management agencies, 1,950 public toilets, 461 convenience service points and 1,303 parking berths, responding to 28.97 million requests from citizens, and receiving 40,571 pieces of public report information. In addition, Hangzhou has relied on the 572 community urban management service offices across the city as well as the smart city management system to establish a "bottom-up" conflict resolution mechanism covering the social, sub-street, district and city levels. This mechanism guides the community urban management service offices to implement the self-location and self-disposal of community problems through the smart urban management system.

Judging from the development trend of China's smart cities, these smart cities are becoming increasingly systematic, integrated, intelligent, and convenient. The information systems of different fields are gradually integrated and the data of various sources form urban big data through sharing and integration. The technologies such as 5G, artificial intelligence and Internet of Things have rapidly become popularized, which has significantly improved the efficiency of urban management and the convenience of urban residents.

Figure 3-24 The Smart Store in Hangzhou offers face scan access and face recognition payment functions (Photo by Xu Kangping)
(Source: http://www.cnsphoto.com)

Chapter 4

Ecological Civilization and the Urban Environment

Ecological progress

Optimization of Ambient Air Quality

Optimization of Water Environment Quality

Soil Environmental Quality Remediation

China Program: Ecological Restoration and City Betterment

>> 4

Ecological Civilization and the Urban Environment

    With the continuous social and economic development of China in the past decades, Chinese cities have witnessed earth-shaking changes, the material and cultural needs of the public have been greatly satisfied, and increasingly strict requirements have been imposed on the promotion of ecological progress. The environmental quality of Chinese cities has undergone wave-like changes with the reform and opening up, and displayed an evolution from good to deterioration to overall improvement. Guided by Xi Jinping's Thought on Socialism with Chinese Characteristics for a New Era, China has continuously promoted the ecological progress, established and practiced the concept that lucid waters and lush mountains are invaluable assets, created a national park system, promoted the central inspections on environmental protection, maintained the ecological security by efforts in biodiversity conservation and vegetation and forest protection, and rationally utilized the clean energy to respond to climate change and carbon emission reduction. The atmospheric, water and soil environment has been gradually improved; the number of days with the urban air quality meeting the environmental standards has risen; the water quality of major rivers and lakes has improved; the urban black and odorous water has been gradually treated, and soil pollution accidents have been effectively controlled and handled. At the same time, in order to deal with "urban diseases", the Chinese government has proposed an "ecological restoration and city betterment" program, and actively promoted the program in 58 pilot cities by three batches nationwide in order to explore more replicable and scalable experience and better improve the living environment from the perspective of comprehensively improving the quality of the urban ecological environment. In short, at present, China has now entered a stage in which the concepts of environmental protection have been more clearly stated, policies and regulations have been continuously improved, and environmental quality is continuously improved, laying a good foundation for building a beautiful China.

4.1 Ecological progress

4.1.1 Status of Ecological Progress in the Past 40 Years of Reform and Opening Up

The 40 years of reform and opening up to the outside world of China has also been the period of China's economic take-off and wave-like changes in environmental quality. With the reform and opening up, the environmental quality has undergone an evolution from good to bad to overall improvement and has now entered a stage where the concepts of environmental protection have been more clearly stated, relevant policies and regulations have been continuously improved, and, therefore, the environmental quality continues to improve.

(1) Deepening the concepts of environmental protection

China has adhered to the concepts of environmental protection synchronized with international concepts, and has become an advocate and promoter of global ecological progress. Environmental protection is the earliest area that was promoted in line with international standards in China, and such concepts as clean production, sustainable development, and circular economy have all been introduced and actively promoted by experts and managers in the field of environmental protection.

(2) Proposing Environmental Protection Policies

The basic principles and policies of environmental protection are based on the national conditions of China, and eight national environmental protection conferences have been held so far. The basic guidelines for effective response have been put forward for the problems existing in different periods.

(3) Improving Environmental Protection Laws and Regulations

China's environmental protection laws and regu-

List of Environmental Protection Concepts in Different Periods Table 4-1

Period	Description of Concepts
1983	Environmental protection was determined as a basic national policy
1994	The 16th executive meeting of the State Council discussed and adopted *China's Agenda 21*, and identified "sustainable development" as a major national development strategy
2003	The ideas of establishing a scientific outlook on development and building a harmonious society were established
2012	Building a Beautiful China became one of the important goals of national development
2017	Building an eco-civilization was laid down as a cornerstone for sustainable development of the Chinese nation. It was proposed to accelerate the reform of the eco-civilization system and build a Beautiful China

(Source: Author's self-painting)

lations have been continuously improved. At present, China has become one of the countries with the most complete environmental regulations in the world and the Chinese government has continuously increased the enforcement of environmental protection. The central inspections of environmental protection cover 31 provinces, autonomous regions and municipalities and the outstanding environmental problems that have concerned the people have been resolved. Accountability has been intensified towards those who fail to implement the decisions and plans adopted by the CPC Central Committee in a resolute or thorough manner or who engage in behaviours, inaction, or acts of abuse (etc.) that engender the destruction of the ecosystem.

(4) Promoting Public Participation

Continuous public participation and positive actions: In 1993, the Trans-Century Cross-China Environmental Protection Tour was launched, and became an important way for public participation and environmental informa-

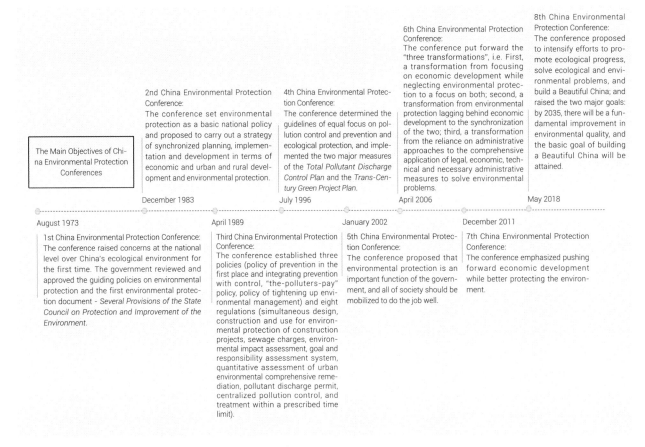

Figure 4-1 Main Objectives of China Environmental Protection Conferences
(Source: Author's self-painting)

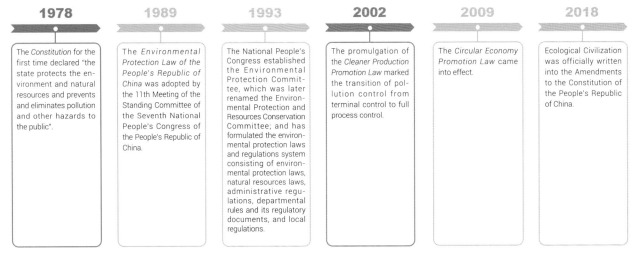

Figure 4-2 Important Timeline for Environmental Protection Related Legislation in the Past 40 Years of Reform and Opening Up
(Source: Author's self-painting)

tion disclosure in China. The bottom anti-seepage project of Yuanmingyuan Lake in 2005 became a landmark event for the public to participate in the decision-making of environmental protection initiatives. Green consumption has become a priority for residents. With the awareness of eco-environmental protection becoming more and more popular, ecological awareness has risen to the consciousness of the entire population in terms of conscious action, how it starts from oneself and one small environmental action. Represented by the Clean Your Plate Campaign, through carrying out CD-ROM operations, promotion of energy-saving home appliances, recycling resources, reducing the use of disposable dinnerware, etc., green, low-carbon, civilized, healthy lifestyles and consumer practices have started to be formed. More and more ordinary volunteers are abandoning the bad habits of over-consumption or throwing things away after their first use, and they are starting to recycle their waste. Green consumption and low-carbon travel have become residents' conscious choices. The environmental protection-oriented social organizations and volunteer teams have been regulated for healthy development, and they have carried out activities according to the law such as public interest litigations for public environmental protection. The social atmosphere of resource conservation and environmental protection has taken initial shape.

4.1.2 Characteristics of Ecological Progress in the New Era: the Report of the 19th CPC National Congress, and the Concept that Lucid Waters and Lush Mountains are Invaluable Assets

The ecological progress will benefit both current and future generations. Recent years have witnessed that the quality of China's ecological environment has continued to improve, and there has been a steady trend towards further improvement. However, due to such factors as the late start of environmental protection efforts, uneven regional and industrial development, and large differences in environmental protection infrastructure, the results already achieved are not stable. In order to push China's ecological progress to a new level, the report of the 19th National Congress of the Communist Party of China laid down ecological progress as a cornerstone for sustainable development for the Chinese nation, regarding it necessary to establish and practice the concept that lucid waters and lush mountains are invaluable assets. The report pointed out that accelerating the reform of the ecological civilization system and building a Beautiful China has set a milestone for the ecological progress in the new era of socialism with Chinese characteristics and provides a fundamental principle and action guide for promoting the new pattern of modernization development characterized by the harmonious development of human kind, nature and for building a Beautiful China.

4.1.3 Ecological Environment Improvement and Management: National Park System, and National Environmental Protection Inspection

(1) Establishing a National Park System

The establishment of a national park system is an important part of China's ecological civilization system. The Chinese government has put forward specific requirements for the establishment of a national park system, emphasizing that "The protection of important ecosystems will be strengthened to ensure their sustainable use. The system of departments independently setting up their own nature reserves, historical and scenic sites, cultural and natural heritage sites, geological parks, and forest parks will be reformed." Furthermore,

Box4-1 Lucid Waters and Lush Mountains are Invaluable Assets

Huzhou, the birthplace of the concept that lucid waters and lush mountains are invaluable assets, the cradle of China's construction of beautiful villages, and the forerunner of green development, uses concrete actions to explain that "lucid waters and lush mountains are invaluable assets." By integrating all aspects if this concept into the whole process of economic, political, cultural and social development, Huzhou has been more resolute in building itself as a firm model. In the past five years, Huzhou's GDP has grown at an average annual rate of 8.4%. In 2017, the growth rates of a total of 23 major economic indicators of Huzhou entered the top four in Zhejiang Province; the total local fiscal revenues exceeded 40 billion yuan for the first time; and the per capita disposable income of urban and rural residents reached 49,934 yuan and 28,999 yuan, up 9.0% and 9.4% respectively. Huzhou has already obtained honors such as being the national civilized city, the national environmental protection model city, the national garden city, and the national forest city.

The megalithic monument bearing the worlds "the lucid waters and lush mountains are invaluable assets" at the entrance of Yucun Village, Tianhuangping Town, Anji County (Photo by He Jiangyong)
(Source: http://www.cnsphoto.com)

Box4-2　Five Waters Treatment Plan in Zhejiang

During the 10-year practice, Zhejiang has regarded the Five Waters Treatment plan as the breakthrough and entry point for implementing the concept that "lucid waters and lush mountains are invaluable assets". The city has regarded the lucid waters and lush mountains as the most generalized welfare for people's life and the fairest public product, continuously improving the quality of the ecological environment. The Five Water Treatment plan of Zhejiang has identified five key priority areas for comprehensive treatment: treating sewage, preventing floods, fixing the drainage system, ensuring the water supply and promoting water conservation. During this period, Zhejiang rectified 4,860 kilometers of rivers with black and odorous water; rectified and eliminated 64,000 enterprises (workshops) with problems; and presented a number of models of rectified and improved massive industries characterized by low efficiency, small scale and decentralized operation including such as Changxing Lead Storage Battery, Pujiang Crystal, Wenling Shoes, and Zhili Children's Wear, etc. The ecological dividend and ecological concepts triggered by the scientific theory that "lucid waters and lush mountains are invaluable assets" has produced strong positive energy in Zhejiang.

The Results of Water Control in Suxi River, Yiwu City, Zhejiang Province (Photo by Lv Bin)
(Source: http://www.cnsphoto.com)

efforts shall be made to "protect the authenticity and integrity of the natural ecological environment and natural and cultural heritage". In September 2017, China issued the *Overall Plan for Establishing a National Park System*, establishing ten pilot national parks including Sanjiangyuan National Park in Qinghai, Shennongjia National Park in Hubei, Pudacuo National Park in Yunnan, Mount Nanshan National Park in Hunan, Mount Wuyi National Park in Fujian, Qianjiangyuan National Park in Zhejiang, Great Wall National Park in Beijing, Mount Qilian National Park, Giant Panda National Park, and the National Park for Siberian Tigers and Siberian Leopards.

Box4-3 Sanjiangyuan National Park

The Pilot Project of Sanjiangyuan National Park is the first pilot project of the national park system approved in China, and with an area of 123,100 square kilometers, it is the biggest in the current list of pilot national parks. Sanjiangyuan is the source of the Yangtze, Yellow and Lancang (Mekong) Rivers. As the "Chinese Water Tower", Sanjiangyuan is an important freshwater supply area in China, with an annual average water output of 60 billion cubic meters, covering more than 600 million people. Sanjiangyuan is the lifeline of water's ecological security in China and even throughout Asia. It is one of the most sensitive regions for global climate change response. It is also one of China's 32 biodiversity priority areas, with 2, 238 species of wild vascular plants and 69 species of key nationally protected wild animals, accounting for 26.8% of the key nationally protected wild animals.

Lakes in the Sanjiangyuan area (Photo by Sun Rui)
(Source: http://www.cnsphoto.com)

(2) National Environmental Protection Inspection

At the end of 2017, the Chinese government completed the first round of central inspections on environmental protection, achieved full coverage inspections of 31 provinces, autonomous regions and municipalities across China, and solved a large number of ecological and environmental problems involving municipal solid waste (MSW), fumes, foul odors, noises, enterprise pollution and black and odorous water as well as other issues. The first round of inspections accepted more than 135,000 complaints from the public, and imposed a total of 29,000 penalties and fines totally 1.43 billion yuan; 1518 cases were investigated and 1,527 persons were detained; 18,448 party and government leading cadres were interviewed, and 18,199 were held accountable for the cases reported.

(3) Environmental Improvement of Water Sources

The Chinese government has continued to promote the remediation of environmental problems in water sources and the goal is to complete the investigation and remediation tasks in the water source areas of the county-level and above cities in the Yangtze River Economic Belt as well as the water source areas of other provinces and municipalities before the end of 2018, involving 6,251 environmental violations of 1,586 water sources in 276 prefecture-level cities from 31 provinces (autonomous regions and municipalities). Through clean-up and remediation, a large number of environmental problems affecting the security of water sources have been effectively solved and the provinces (autonomous regions and municipalities) have closed down and banned 180 sewage outlets in the water source protection areas, rectified and solved 792 pollution problems from industrial enterprises, effectively renovated 686 cases of dock and road traffic-crossing problems, and promoted the resolution of 2,260 rural non-point source problems.

4.1.4 Ecological Security: Biodiversity, and Vegetation Protection

(1) Biodiversity Conservation

China is one of the most biologically diverse countries in the world and one of the first countries to join the *Convention on Biological Diversity*. In 2017, the Chinese government launched a number of major biodiversity conservation projects to carry out field rescue and reproduction of rare and endangered wild plants of very small populations, and in so doing established more than 440 biodiversity observation plots. In 2017, with the newly-added 17 new national-level nature reserves, a total of 463 national nature reserves have been established in China. The issuance of regional land-reclamation quotas for the year 2017 was suspended, and special inspections on land-reclamation were carried out in 11 coastal provinces (autonomous regions and municipalities). Efforts are being made to promote ecological restoration projects such as "Blue Bay" and "Ecological Island Reef", rectify more than 70 kilometers of shoreline and repair more than 2,100 hectares of coastal wetlands. In May 2018, China released the *Redlist of China's Biodiversity — Macrofungi* and *Catalogue of Life China 2018*.

(2) Special Supervision and Inspection over National Nature Reserve

The Chinese government launched a special campaign for the supervision and inspection of the "Green Shield 2017" on national nature reserves. This action was a special action with the widest scope of inspections, the largest number of cases under investigation, the greatest efforts ever invested in rectification, and the most stringent pursuit of accountability since the establishment of nature reserves in China. By the end of 2017, a total of 20,800 issue indicators involving 446 national nature reserves were investigated, more than 2,460 illegal enterprises were shut down, and more than 5.9 million square meters of unauthorized structures and facilities were demolished; more than 1,100 people were held accountable, including 60 department-level governmental officials and 240 division-level officials. More than 13,100 rectifications were completed, accounting for 63% of the total number of problems identified. A record cataloguing the remaining issues has been established and these problems are being rectified.

(3) Vegetation and Forest Protection

The eighth national forest inventory data show that China's forest area is 208 million hectares, the forest reserves are 15.137 billion cubic meters, and the forest coverage rate is 21.66%, making China the country with the fastest growing forest resources in the world. Among them, the artificial forest area is 69.33 million hectares, accounting for 36% of China's area of afforested land; and the stock volume of the artificial forests is 2.483 billion cubic meters, accounting for 17% of China's stock volume of forests. China's artificial forests rank top in the world in terms of size.

4.1.5 Resource Security: Resource Protection and Energy Utilization

(1) Natural Resource Protection

The Chinese government has set up the Ministry of Natural Resources to uniformly exercise the stewardship duties of all common natural resources, their assets belonging to all people. This body uniformly exercises responsibilities for the utilization and control of all state land space and ecological protection and restoration, focuses on solving problems such as unsatisfactory individual stewardship of natural resources and the overlapping of spatial planning, and realizes the overall protection, systematic restoration and comprehensive management of mountains, waters, forests, lakes and grasslands. The establishment of the Ministry of Natural Resources marks China's entry into the era of resource management.

(2) Utilization of Water Resources

In order to vigorously promote water conservation among the public and ensure national water security, the Chinese government has issued the *National Action Plan for Water Conservation* to determine 29 specific tasks with the target that by 2035, the total water consumption in China will be controlled within 700 billion cubic meters, and the resource conservation and recycling of water will reach the world's advanced level. According to the China Water Resources Bulletin 2017, the total amount of water resources in China was 2,876.12 billion cubic meters in 2017, which was 3.8% more than the average of previous years;

Figure 4-3 Voluntary Tree Planting Activities in Spring of 2019 Held in Xiong'an New District of Hebei Province (Photo by Han Bing)
(Source: http://www.cnsphoto.com)

the national water supply was 604.34 billion cubic meters, which was 320 million cubic meters more than the total water supply in 2016; the national per capita comprehensive water consumption is 436 cubic meters; the water consumption per 10,000 yuan of GDP (at current prices) is 73 cubic meters; the urban domestic water consumption per capita is 22 liters/day; and the rural domestic water consumption per capita is 87 liters/day.

(3) Clean Energy Utilization

China's energy structure is shifting from coal-based to diversified, and the energy development momentum is shifting from traditional energy growth to new energy growth. According to data released by the National Energy Administration, by the end of 2017, the total installed capacity of power generation nationwide totaled 1.78 billion kilowatts, and the installed capacity of renewable energy power generation reached about 650 million kilowatts. Since the 18th National Congress of the Communist Party of China, the proportion of coal consumption has dropped by 8.5 percentage points (pp), and the proportion of clean energy consumption has increased significantly. In 2017, the proportion of non-fossil energy and natural gas consumption reached 13.8% and 7% respectively, an increase of 4.1pp and 2.2 pp respectively; the amount of electric energy substitution reached more than 100 billion kWh, the amount of natural gas substitution reached 30 billion cubic meters; and the output of natural gas was about 150 billion cubic meters, with China's global ranking rising from 18th to 6th place.

(4) Responding to Climate Change and Carbon Emission Reduction

After years of efforts, China has initially reversed the rapid growth of greenhouse gas emissions with its growth rate slowing down significantly. According to the *China's Policies and Actions for Addressing Climate Change (2018)*, in 2017, China's carbon dioxide emissions per unit of gross domestic product (GDP) (hereinafter referred to carbon intensity) declined by approximately 46% compared to 2005, already exceeding the 2020 target of reducing carbon intensity by 40%-45%. Non-fossil fuel energy accounted for 13.8% of primary energy consumption and is expected to

Box4-4　Beijing Plain Area Basically Converted All Coal-fired Boilers into Gas-fired Operations

In 2013, Beijing officially launched the special action in rural areas titled "Reducing Coal Consumption and Replacing Coal to Clean the Air" and achieved the goal of fundamental "coal-free" operation in the urban area and villages of the four districts in south Beijing by 2017, and completed the task of "replacing coal with clean energy" among the village residents of 450 rural areas. At present, the Beijing plain area has essentially converted all coal-fired boilers into gas-fired operations. About 1,110,000 households in 2,963 villages have bid farewell to coal-fired heating, of which 858,100 households in 2,279 villages have implemented "coal-to-clean energy" conversion (coal-to-electricity conversion for 1,822 villages, accounting for 80%; "coal to gas" conversion for 457 villages, accounting for 20%), and another 250,000 households in 684 villages have completed demolition and moved into high-rise residential buildings, and mainly adopted centralized clean-energy-fired heating.

reach the target of 15% by 2020. The forest stock has increased by 2.1 billion cubic meters and has exceeded its target for 2020. At the end of 2017, the *National Carbon Emissions Trading Market Construction Plan (Power Generation Industry)* was released, which launched the China Carbon Emissions Trading System and launched a pilot project on carbon emissions trading in seven provinces and cities. As of November 2018, the transaction volume reached 270 million tons of carbon dioxide, and the turnover exceeded 6 billion yuan.

4.2 Optimization of Ambient Air Quality

4.2.1 Overall Condition of the Atmospheric Environment

(1) Air Quality

According to the data of *China Ecological Environment Statement 2018*, in 2018, among the 338 prefecture-level and above cities, 121 cities have met environmental air quality standards, accounting for 35.8% of the total number of cities, up 6.5 pp over 2017; The ambient air quality of 217 cities exceeded the standard, accounting for 64.2%, down 6.5 pp from 2017. 338 cities recorded an average of 79.3% of days with good air quality, up by 1.3 pp from 2017; 338 cities suffered 1899 heavily polluted days during daylight hours, 412 fewer days than 2017, and 822 severely polluted days during daylight hours: the number of days with PM2.5 as the primary pollutant accounted for 60.0% of the number of those heavily polluted (or worse) days, 37.2% for the number of days with PM10 as the primary pollutants, and 3.6% for the number of days with O_3 as the primary pollutant.

Figure 4-4 Proportion of ambient air quality levels in 338 cities in 2018
(Source: China Ecological Environment Statement 2018)

(2) Pollutant Emissions

The concentrations of PM2.5, PM10, O_3, SO_2, NO_2 and CO were 39 μg/m³, 71 μg/m³, 151 μg/m³, 14 μg/m³, 29 μg/m³ and 1.5 mg/m³, respectively, with the ratio of the number of days exceeding the standard in 2018 reaching 9.4%, 6.0%, 8.4%, less than 0.1%, 1.2% and 0.1% respectively. Compared with 2017, the O_3 concentration and the ratio of the number of days exceeding the O_3 standard both increased, while the concentration and the ratio of the number of days exceeding the standard of the other five indicators all decreased.

Without deducting the influence of sand and dust, the proportion of cities meeting the air quality standards reached 33.7% among 338 cities, and the proportion of cities exceeding the air quality standards was 66.3%; the average concentrations of PM2.5 and PM10 were 41 μg/m³ and 78 μg/m³, respectively, down by 6.8% and 2.5% respectively compared to 2017.

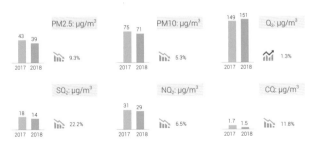

Figure 4-5 Inter-annual Variations of the Concentrations of Six Pollutants in 338 cities in 2018
(Source: China Ecological Environment Statement 2018)

Figure 4-6 Inter-annual Variations of the Proportions of Days Exceeding the Standards of Six Pollutants in 338 cities in 2018
(Source: China Ecological Environment Statement 2018)

(3) Acid Rain

In 2018, 5.5% of China's land area, i.e. about 530,000 square kilometers, suffered from acid rain, down by 0.9 pp from 2017. Among them, the area with moderately heavy acid rain accounted for 0.6% of China's land area. Acid rain pollution is mainly distributed in the south of the Yangtze River and east of the Yunnan-Guizhou Plateau, mainly including most parts of Zhejiang, Shanghai, northern Fujian, central Jiangxi, central and eastern Hunan, central Guangdong and southern Chongqing. The annual average pH value of nationwide precipitation ranges from 4.34 (Dazu District, Chongqing) to 8.24 (Kashgar City, Xinjiang), with an average of 5.58. The proportions of cities suffering from acid rain, moderately heavy acid rain and heavy acid rain were 18.9%, 4.9% and 0.4% respectively.

Among the 471 cities (districts and counties) that

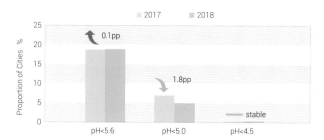

Figure 4-7 Inter-annual Variations of the Proportion of Cities with Different Annual Average pH Values of Precipitation in 2018
(Source: China Ecological Environment Statement 2018)

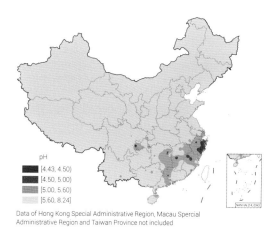

Figure 4-8 Schematic Diagram of Contour Distribution of the Annual Average pH Value of Precipitation in China in 2018
(Source: China Ecological Environment Statement 2018)

monitor their precipitation, the average frequency of acid rain is 10.5%, down by 0.3 pp from that of 2017. The proportion of cities suffering from acid rain was 37.6%, up by 1.5 pp from that of 2017; and the proportions of cities with acid rain frequency of 25% and above, 50% and above and 75% and above were 16.3%, 8.3% and 3.0% respectively.

4.2.2 Distribution and Characteristics of Haze

In 2017, the national atmospheric environment was further improved. According to the official data of the China Meteorological Administration, there were 6 large-scale haze weather process and 9 sand-dust weather processes in China during the year, all of which were less than normal. The national average number of hazy days was 27.5 days, a decrease of 10.5 days from 2016 and a decrease of 19.4 days from 2013. The average numbers of hazy days in Beijing-Tianjin-Hebei region and the Yangtze River Delta in 2017 was 42.3 days, 53.3 days, and 17.9 days, respectively, registering a decrease of 18.1 days, 17.6 days, and 3.2 days from 2016 and a decrease of 28.8 days, 35.7 days, and 15.6

days from 2013. The total amount of atmospheric nitrogen dioxide and sulfur dioxide continued to decline throughout the year.

4.2.3 Haze Management—Implementation of National 10-chapter Air Pollution Prevention and Control Action Plan and the Blue Sky Defense War

According to the *National Air Pollution Prevention and Control Work Progress and Recommendations*, the proportion of primary coal use in primary energy consumption in 2017 decreased by more than 8 pp from 2012, and the proportion of coal in primary energy consumption decreased from 67.4% to 60.3%. In the urban developed area, more than 200,000 small coal-fired boilers with a capacity smaller than 10 ton/hour and no hope for renovation were eliminated. The coal-fired units with the total capacity of 700 million kilowatts nationwide have completed ultra-low emission conversion, accounting for 71% of the installed capacity of coal-fired power plants. Local governments eliminated outdated production capacity and addressed the overcapacity of over 200 million tons of steel, 250 million tons of cement, 110 million weight cases of sheet glass, 25 million kilowatts of coal-fired power units, etc., and the substandard steel production capacity of 140 million tons was cleared.

In June 2018, the State Council issued the *Three-Year Action Plan for Winning the Blue Sky Defense War (2018-2020)*, which requires that by 2020, the total emissions of sulfur dioxide and nitrogen oxides be reduced by more than 15% from 2015; the PM2.5 concentration of cities at the prefecture level and above that fail to meet the standard will be decreased by more than 18% compared with 2015. The ratio of days with superior air quality in cities at the prefecture level and above will reach 80%, and the ratio of days with severe and worse pollutions shall decrease by more than 25% compared with 2015.

4.3 Optimization of Water Environment Quality

4.3.1 Overall Status of Water Environment Quality

According to the date of the *China Ecological Environment Statement 2018*, among the 1935 monitoring sections (point location) under national surface water monitoring, in 2018 the proportion of sections with Grade I to III water quality was 71.0%, an increase of 3.1 pp higher than that of 2017; and the proportion of the sections with worse than Grade V water quality was 6.7%, which is 1.6 pp lower than that of 2017.

In 2018, among the 1,613 water quality sections monitored in the seven drainage basins of Yangtze River, the Yellow River, the Pearl River, the Songhua River, the Huaihe River, the Haihe River, the Liaohe River, the Zhejiang-Fujian regional rivers, the Northwest China rivers, and the Southwest China rivers, Grade I sections accounted for 5.0%, Grade II sections accounted for 43.0%, Grade III sections accounted for 26.3%, Grade IV sections accounted for 14.3%, Grade V sections

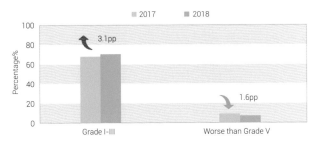

Figure 4-9 Inter-annual Variations of Surface Water Quality Grades Nationwide in 2018
(Source: China Ecological Environment Statement 2018)

accounted for 4.5%, and sections worse than Grade V accounted for 6.9%. Compared with 2017, the proportion of Grade I water quality sections increased by 2.8 pp, that of Grade II sections increased by 6.3 pp, that of Grade III sections decreased by 6.6 pp, that of Grade IV sections decreased by 0.2 pp, that of Class V sections decreased by 0.7 pp, and that of sections worse than Grades V decreased by 1.5 points pp.

In 2018, among the 111 important lakes (reservoirs) that monitored water quality, there were 7 lakes (reservoirs) with Grade I water quality, accounting for 6.3% of the total; 34 with Grade II water quality, accounting for 30.6%; 33 with Grade III water quality, accounting for 29.8%; 19 with Grade IV water quality, accounting for 17.1%; 9 with Grade V water quality, accounting for 8.1%; and 9 with worse than Grade V water quality, accounting for 8.1%. The main pollution indicators are total phosphorus, chemical oxygen demand and permanganate index. Among the 107 lakes (reservoirs) that monitored the trophic status, 10 were in the oligotrophic state, accounting for 9.3%; 66 were in the mesotrophic state, accounting for 61.7%; 25 were in the mildly eutrophic state, accounting for 23.4%; and 6 were in the moderately eutrophic state, accounting for 5.6%.

Figure 4-10 The Overall Water Quality of the Nationwide Drainage Basins in 2018
(Source: China Ecological Environment Statement 2018)

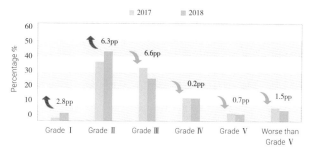

Figure 4-11 Inter-annual Variations of Overall Water Quality in the Nationwide Drainage Basins in 2018
(Source: China Ecological Environment Statement 2018)

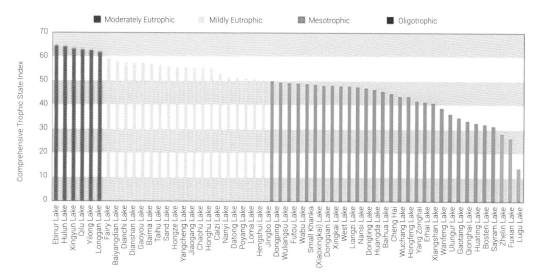

Figure 4-12 Comparison of Trophic State of Important Lakes in 2018
(Source: China Ecological Environment Statement 2018)

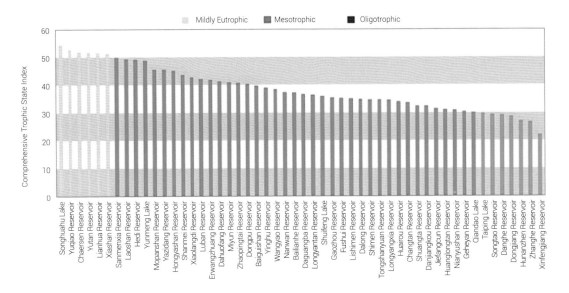

Figure 4-13 Comparison of trophic state of important reservoirs in 2018
(Source: China Ecological Environment Statement 2018)

4.3.2 Major Water Pollution Accidents and Pollution Control and Treatment

According to data released by the Ministry of Ecology and Environment, in 2017, there were 302 environmental emergencies in China, a decrease of 0.7% compared with 2016. Among them, there was 1 severe event, a decrease of 2 events on a comparative annual basis; 6 large events, an increase of 1 case on a comparative annual basis; and the rest were general events. The severe environmental emergency was the thallium pollution in the Guangyuan section, Sichuan of Jialing River caused by waste water discharge from Hanzhong Zinc Industry Copper Mine Co., Ningqiang County, Shaanxi Province.

Box4-5 Water Pollution Incidents and Treatment in Jialing River

On May 5, 2017, a thallium pollution incident occurred in the Guangyuan section of the Jialing River, resulting in the thallium concentration exceeding the standard by 4.6 times in the potable water source for Xiwan Water Plant. The People's Government of Guangyuan City initiated the Class II contingency response, set up an emergency headquarters, deployed emergency response work such as an emergency water supply, storage adjustment for dilution, and water plant processing reform, jointly carried out pollution source investigations with the Hanzhong City Government of Shaanxi Province on the main stream and tributary channels along the Jialing River, and locked down the enterprise perpetrating the pollution through sample monitoring, scientific analysis and other measures. The characteristics of this incident include: strengthened early warning and first moment response; regional linked operations and joint movements; scientific analysis, rapid and orderly disposal; timely disclosure of information, and responding to public concerns.

4.3.3 Water Environment Control: Target Requirements of the 10-Chapter Water Pollution Prevention and Control Plan, River Chief System

In April 2015, the State Council issued the *Notice of the State Council on Issuing the Action Plan for Prevention and Control of Water Pollution* (10-chapter water pollution prevention and control plan), requiring that by 2020, the overall proportions of the water quality of seven basins, including the Yangtze River basin, the Yellow River basin, the Pearl River basin, the Songhua River basin, the Huaihe River basin, the Haihe River basin and the Liao River basin to be as follows: above average quality (reaching or exceeding Grade III) to be 70% or above; the quantity of black and odorous water bodies in built-up areas in cities at prefecture level and above to be controlled within 10%; the overall proportion of centralized drinking water source quality in cities at prefecture level and above reaching or exceeding Grade III to be larger than 93%; the proportion of extremely poor groundwater quality nationwide to be controlled around 15%; and the proportion of above average (Grade I and II) water quality in offshore areas to reach about 70%. The proportion of unusable (worse than Grade V) water sections in Beijing-Tianjin-Hebei Region is required to be lowered by about 15%, and efforts should be made to eliminate unusable water bodies in the Yangtze River Delta and Pearl River Delta. By 2030, the overall proportion of water quality in seven key basins nationwide reaching above average will reach 75% or above, with black and odorous water bodies in urban built-up areas generally eliminated and essentially 95% of urban centralized drinking water will have a source quality reaching or exceeding Grade III.

In November 2016, the Central Office and the State Council issued the *Opinions on Comprehensively Promoting the River Chief System*. By the end of June 2018, 31 provinces, autonomous regions, and municipalities directly under the Central Government had fully established a river chief system, and more than 300,000 river chiefs for all rivers in the 31 provinces, autonomous regions, and municipalities directly under the Central Government are clearly specified at the provincial, city, county and township levels, of which 402 river chiefs are provincial leaders. The local governments have established six systems including the river chiefs' meeting system, information sharing system, information reporting system, work supervision system, assessment accountability and incentive system, and inspection and acceptance system. By July 2018, provincial level river chiefs have inspected the rivers or lakes for 926 person-times, and the river chiefs at the city, county and township levels have inspected the rivers or lakes for more than 2.10 million person-times.

4.3.4 Urban water ecological environment construction

In October 2018, the *Implementation Plan for the War on Urban Black and Smelly Water Body Control* issued upon the approval of the State Council further clearly stated that "by the end of 2018, over 90% of black and odorous water bodies in the municipalities directly under the central government, provincial capital cities, and cities under separate state planning shall be eliminated to achieve long-term clear waters under long-term mechanism. By the end of 2019, the proportion of black and odorous water bodies eliminated in other prefecture-level urban built-up areas will be significantly improved, reaching more than 90% by the end of 2020. The urban built-up areas in Beijing-Tianjin-Hebei region, Yangtze River Delta, and Pearl River Delta region shall be encouraged to eliminate black and odorous water bodies as soon as possible."

By the end of 2018, the total number of black and odorous water bodies identified in China reached 2,100. Among them, there were 1,745 black and odorous water bodies that had completed treatment; 264 water bodies under treatment; and 91 water bodies in the process of plan formulation. The length of the black and odorous water bodies that had completed treatment is 6970.66 kilometers, covering an area of 1279.28 square kilometers and accounting for 87% of the total.

In October 2018 and May 2019, experts from the Ministry of Housing and Urban-Rural Development, the Ministry of Ecology and Environment, and the Ministry of Finance determined the lists of 40 black and odorous water treatment demonstration cities in 2 batches for 2018 and 2019 through on-site defense and reviews. Through the implementation of urban black and odorous water treatment, the quality of urban water ecological environment in China is gradually improving. The efforts for the next stage shall focus on the promotion of the rectification of black and odorous waters in key cities that have not reached the governance targets and cities above the prefecture level in the Yangtze River Economic Belt.

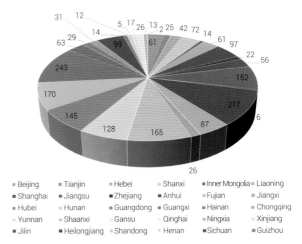

Figure 4-14 The number of black and odorous water bodies identified in each province (not including HongKong, Macao or Taiwan)
(Source: http://www.hcstzz.com)

Box4-6 Rectification of Beijing Yudai River

Yudai River starts in Liyuan South Street, Tongzhou District, Beijing, and the area along the river is densely populated. Due to the direct discharge of sewage and the imperfect sewage interception pipeline, the water in the Yudai River is black and odorous, which discouraged the surrounding residents. In 2017, Tongzhou District Government carried out the renovation project of Yudai River. Unlike other black and odorous water treatments, the Yudai River remediation plan paid more attention to the comprehensive functions of the natural ecosystem of the urban water body including water landscape construction and ecological restoration, and the flood storage and detention areas. The project highlighted cultural heritage, dug deep, protected and passed on the historical and cultural resources centered on the Grand Canal, and transformed and restored historical sites including the ancient river course and 10 ancient wharfs of the Yudai River to fully reflect the Chinese cultural genes.

Box4-7 Significant achievements in urban water ecological environment improvements

Looking back on the past 40 years, the evolution of China's urban water environment has gone through three stages and has endured the consequences and the black stink of the extensive development, pursued improvements through governance, expected the harmony of ecological development, and returned to the state where fishes swim in the clear waters.

From 1978 to 1998, after experiencing extensive and rapid development in the early stage of reform and opening up, the problem of water pollution in China gradually appeared. The quality of surface waters deteriorated and pollution increased, which was reflected in the gradual serious organic pollution and eutrophication of lakes and urban water bodies had turned from clean to black and odorous..

From 1998 to 2012, under the influence of the "Action Zero Hour" for pollution control in the Huaihe River, the urban water environment opened the era of comprehensive governance. However, due to excessive cumulative burdens from the previous period, the water environment management in this period has had little effect, and urban water environment problems are frequent.

From 2013 to 2019, after the Central Urbanization Work Conference, the urban water environment treatment was mainly centered on protection and restoration. Under the guidance of relevant national policies on the sewage interception pipeline, the construction of new sewage treatment plants and upgrading and renovation of existing sewage treatment plants as well as the construction of sponge cities, the treatment of black and odorous water bodies, urban water environment management has experienced from "point to surface" development, i.e. the systematic governance from single plant to the integration of the treatment plants with networks and the integration of treatment plants with networks and river (lake) banks and the comprehensive management with the focus from sewage interception and control to focus on ecological restoration. After a series of treatments, the overall quality of urban waters in China is currently improving, and the people's demand for a better living environment is generally satisfied.

4.4 Soil Environmental Quality Remediation

4.4.1 Overall status of soil environmental quality

According to the data of *China Ecological Environment Statement 2018*, as of the end of 2017, China had a total of 644.864 million hectares of agricultural land, including 134.881 million hectares of arable land, 14.214 million hectares of garden plots, 252.802 million hectares of forest land, 219.320 million hectares of grassland, and 39.574 million hectares of construction land, including 32.131 million hectares of urban villages and industrial and mining land. According to the data of *China Ecological Environment Statement 2017*, the average quality grade of cultivated land in China is 5.09, among which the area of cultivated land of Grade one to three is 555 million *mu*, accounting for 27.4% of the total cultivated land; the area of cultivated land of Grade four to six is 912 million *mu*, accounting for 45.0% of the total cultivated land; the area of cultivated land of Grade seven to ten is 559 million *mu*, accounting for 27.6% of the total cultivated land.

According to the results of the first national census for water, the total land area suffering from soil erosion in China is 2.949 million square kilometers, accounting for 31.1% of the total area covered by the census. Among them, the land area suffering from water erosion is 1.293 million square kilometers, and the land area suffering from wind erosion is 1.656 million square kilometers. According to the results of the 5th National Desertification and Sandification Monitoring, China's desertified land area is 2,611.6 thousand square kilometers, and the sandified land area is 1,721,200 square kilometers.

4.4.2 Soil pollution accidents and pollution control treatment

From April 2005 to December 2013, China conducted the first national survey on soil pollution. According to the *Bulletin of National Soil Pollution Survey* (the first national soil pollution survey in 2014), the overall soil environmental conditions in China are not optimistic, and the total over-standard rate of national soil is 16.1%, of which slight, mild, moderate and severe pollution ratio are 11.2%、2.3%、1.5% and 1.1% respectively. The main type of pollution is inorganic contamination, followed by organic contamination, and the proportion of compound contamination is small. The number of sites with inorganic pollutants exceeding the standard accounts for 82.8% of all over-standard sites. From the perspective of pollution distribution, soil pollution in south region is heavier than that in the north region; soil pollution problems in some areas such as the Yangtze River Delta, the Pearl River Delta, and the old industrial bases in Northeast China are more prominent; the soils in the southwest and south-central regions have a large range of heavy metals; and the distribution of four inorganic pollutants of cadmium, mercury, arsenic, and lead showed a trend

Box4-8 China Fine-chemical (Taixing) Park backfilled hazardous wastes on the inside of the Yangtze River bank

In June 2018, the Fourth Environmental Protection Inspection Team of the CPC Central Committee discovered during the grassroots inspection in Taizhou City, Jiangsu Province that local government of Taixing City of Taizhou was perfunctory in handling the matters delivered to it by the central environmental protection supervision team in July 2016, and China Fine-chemical (Taixing) Park (hereinafter referred to as Taixing Chemical Park) was found to have backfilled more than 30,000 cubic meters of chemical wastes and other solid wastes, including a large amount of hazardous wastes, on the inside of the bank of the Yangtze River. Upon investigation, it was found that the concerned location covered an area of nearly 11,000 square meters and it was determined that the total amount of wastes including retained chemical wastes was 31,950 cubic meters, and the surrounding soil and groundwater had been polluted. The above-mentioned problems have been highly valued by competent authorities of Jiangsu Provincial Government, and relevant responsible persons had been punished.

of gradual increase from northwest to southeast and from northeast to southwest.

4.4.3 Improvement of Soil Environment: Target Requirements of the 10-Chapter Soil Pollution Action Plan

In May 2016, the State Council issued the *Soil Pollution Prevention and Control Action Plan* (10-Chapter Soil Pollution Action Plan), which required that by 2020, the worsening national soil pollution will be initially curbed, the overall soil environmental quality will remain stable, the soil environment security of agricultural land and construction land will be essentially guaranteed and soil environmental risks will, in essence, be put under control. By 2030, the quality of the soil environment in China will be stable with further improvements, and the soil environmental security of agricultural land and construction land will be effectively guaranteed, and the soil environmental risks will be fully controlled. By the middle of this century, the quality of the soil environment will be fully improved, and the ecosystem will achieve a virtuous circle. By 2020, the safe utilization rate of contaminated cultivated land will reach 90%, and the safe utilization rate of contaminated land plots will reach over 90%. By 2030, the safe utilization rate of contaminated cultivated land will reach over 95%, and the safe utilization rate of contaminated land plots will reach over 95%.

4.5 China Program: Ecological Restoration and City Betterment

In order to control "urban diseases" and improve the living environment, the Chinese government has proposed an "ecological restoration and city betterment" program, and fully launched a comprehensive evaluation of urban construction and ecological environment in various cities in 2017, and began to formulate an implementation plan for ecological restoration and city betterment and promote a batch of productive demonstration projects. "Ecological restoration and city betterment" refers to restoring the damaged natural environment and topography of the cities, and improving the quality of the ecological environment with the concepts of re-ecology; and using the concept of and darning to demolishing unauthorized buildings, restoring urban facilities, space environment and landscape, and enhance the urban features and vitality with the concepts of renewal and weaving. As of 2018, the Chinese government has announced three batches of 58 pilot cities for "ecological restoration and city betterment" to explore and summarize more experience that can be replicated and promoted for "ecological restoration and city betterment".

4.5.1 Local practice

Considering their own urban characteristics, the pilot cities promoted supply-side structural reforms to make up the shortcomings of the cities, and promote urban transformation through the "ecological restoration and city betterment" initiative, and have achieved remarkable results. However, at the same time, it is recognized that the "ecological restoration and city betterment" initiative still has a long way to go in the future and it is necessary to continuously improve the various functions of the cities, meet the needs of the citizens, and improve the urban livability, security and resilience, prosperity and vitality, harmony and inclusiveness in an all-round way.

Box4-9 Ecological restoration and city betterment practice in Sanya (first batch)

In 2015, Sanya became the first pilot city in China for "ecological restoration and city betterment" and was scheduled to solve the urban diseases in Sanya based on the comprehensive environmental development and quality improvement of the city. The local government initiated 18 key projects with a total investment of about 2.6 billion yuan, including the ecological restoration projects covering the urban mountains, rivers, seas and wetlands, as well as city betterment projects such as the construction of urban green space and green belts, advertising board management, urban toning change, lighting for urban night scenes, urban skyline planning, urban building façade renovation, and governance of unauthorized structures and buildings. In 2015 and 2016, a total of about 7.7 million square meters of unauthorized structures and buildings were demolished, and the purpose and significance of "ecological restoration and city betterment" was promoted to more than 700,000 permanent residents in Sanya. The city has played a good demonstration role in the ecological restoration and city betterment initiative.

Nanshan Cultural Tourism Area of Sanya, Hainan Province (Photo by Chen Wenwu)
(Source: http://www.cnsphoto.com)

Box4-10 Ecological restoration and city betterment practice in Fuzhou (second batch)

In March 2017, Fuzhou was included in the second batch pilot cities of "ecological restoration and city betterment" initiative. Adhering to the respect for history and culture and inheritance of the cultural context, Fuzhou has created specialty historical and cultural blocks to reproduce the historical landscape pattern and realize the continuation of the urban cultural context and the reconstruction of the park system. Since the launch of the pilot project, 86 of the 107 inland rivers of Fuzhou City have systematically completed the comprehensive improvement, and the water environment has been significantly improved. Three historical areas such as Shangxiahang, Zhuzifang and Yantaishan have been gradually restored and opened to the public, and the renovation and protection of 232 traditional old streets and alleys have been launched and the renovation of 113 streets and alleys have been completed. Among the 415 residential areas undergoing renovation and improvement, 335 old residential areas have road lamps installed, traffic cleared, and more green space added and have become beautiful.

Shangxiahang Historical and Cultural Block after restoration and renovation (Photo by Lv Ming)
(Source: http://www.cnsphoto.com)

Box4-11   Ecological restoration and city betterment practice in Xuzhou (third batch)

In July 2017, Xuzhou City was listed as one of the third batch of pilot cities for ecological restoration and city betterment in China. Xuzhou promotes the transformation and development of resource-based city through "ecological restoration and city betterment", and combines the transformation of urban functions to explore working methods, technical routes and implementation strategies for environmental improvement and urban transformation and development, and explore new paths for green transformation and development of resource-based cities. The local government has carried out ecological restoration and city betterment work according to local conditions, successively rectified more than 90,000 mu of coal mining subsidence land, expanded 35,000 mu of newly cultivated land, and fully completed the comprehensive improvement of 8 rivers such as the Abandoned Yellow River and Kuihe River in the urban area; insisted on the simultaneous renovation of shanty towns and old residential areas, and the total renovation area was 37.9 million square meters, solving the housing problems of nearly 200,000 shanty town households.

4.5.2   Green Eco-district

China has officially launched the construction of green eco-districts for nearly 12 years, and the construction of eco-districts in various places has yielded fruitful results. According to the data of *China Low Carbon Eco-City Development Report 2018*, between 2007 and 2015, China announced 139 new green eco-district projects, and from 2017 to 2018, six new international cooperation projects of China Eco-city Academy were added. After 2017, the development focus of green eco-districts was integrated into the construction requirements of sponge cities and ecological restoration and city betterment, and at the same time highlighted the construction of smart community.

China's green buildings have made tremendous achievements and progress, and have had a positive impact on China and the world. According to the *China Green Building Report 2017*, as of 2018, China's green buildings accounted for 40% of new urban construction. In 2005-2016, the domestic LEED (Leadership in Energy and Environmental Design) certification area has a compound annual growth rate of 77%. As of August 2017, the cumulative certification area has exceeded 48 million square meters, with a footprint over 54 cities. China has become the world's largest LEED certification market country beyond the US.

Box4-12　Nanjing Hexi New Town Green Ecological City

At the beginning of 2012, Hexi New Town began to build a green eco-district. After years of construction practice, it has essentially met the indicators of eco-district construction. In terms of green buildings, Hexi New Town has obtained green building design logo for projects of more than 1 million square meters, and has built a green building area of about 3.5 million square meters, and buildings with two stars and above accounted for 76.2% of the total. As for green transport, the full length of the tram line 1 opened for operation reached 7.76 kilometers with 13 stations and interchangeable transfer with the subway and bus lines, and a real-time, accurate and efficient comprehensive traffic management and travel service system has been established to improve the efficiency of the traffic operation of road network; as for green municipal services, Hexi constructed the Youth Olympic Energy Center to improve the utilization of renewable energy and clean energy; as for green environment, the green coverage and per capita green area have reached 48% and 22 square meters respectively. At present, Hexi New Town is developing towards a demonstration city of regional integration, and is committed to operation and management of utility tunnels, construction of sponge cities, industrialization of buildings, construction of smart cities, and improvement of energy efficiency of green buildings.

Twin Towers of Youth Olympic Center (Nanjing International Youth Cultural Center) in Nanjing Hexi New Town (Photo by Yang Bo)
(Source: http://www.cnsphoto.com)

Chapter 5

Culture City

Cultural Heritage Deeply Rooted in History

People-oriented Public Space

China Program: Industrial Heritage Protection and Reuse

Culture City

Culture is the soul of a country and a nation. A rising culture rejuvenates a country and a strong culture creates a strong nation. Without a high degree of cultural self-confidence or cultural prosperity, there would be no great rejuvenation of the Chinese nation. In the process of rapid urbanization, the development and construction of China's cities are faced with the threat of the homogenized image of cities, and the decline of memories and culture. The protection of human cultural heritage and urban cultural characteristics has entered the most urgent and crucial historical stage. In December 2018, the National Working Conference on Housing and Urban-Rural Development was held in Beijing, proposing to promote the construction of cultural cities, further strengthen the protection of famous towns in famous historical and cultural cities, promote the preservation, utilization, renewal and renovation of existing buildings, improve the urban design system, and strengthen the management of architectural design.

In the globalized landscape, history and culture are an important part of urban soft power, and cultural innovation is an important driving force for urban transformation and sustainable development. In the period of transformation development, each city must be examined from the perspective of cultural strategy; protection and development of cities must be promoted on a reciprocal basis; and efforts must be made to promote the cultural heritage that is deeply rooted in history in the overall development of the cities.

Urban culture is closely related to people's lives, so we must focus on the important shifting of cultural development and urban construction from "material-centered" to a "people-centered" approach. Therefore, promoting the construction of "culture cities" is not only an inheritance and perpetuation of urban culture, but also an important measure to implement these people-oriented construction concepts. On the occasion of the 40th anniversary of reform and opening up, the concepts of planning and construction of cultural cities have taken deep roots in the hearts of the public and inspiring achievements have been achieved nationwide.

5.1 Cultural Heritage Deeply Rooted in History

In 2017, in the report to the 19th CPC National Congress, it was proposed that we should "promote the creative transformation and innovative development of Chinese excellent traditional culture" and "enhance the protection and utilization of cultural relics and the protection and inheritance of cultural heritages". During his investigation in Guangzhou in October 2018, when talking about urban planning and construction, CPC General Secretary Xi Jinping emphasized that "the inheritance of civilization and the continuation of in cities are very important. The traditional and modern aspects of cities shall be integrated for development, and allow the cities to preserve their memories letting people reminisce about nostalgia."

Since the reform and opening up, China's historical and cultural heritage protection regulations as well as systems have been continuously improved upon. Especially since the establishment of the system a state-list for protection of famous historical and cultural cities, towns & villages in 1982, China has protected a large number of precious historical and cultural heritages in the process of rapid urbanization and has played an important role in the continuing historical cultural context, protecting cultural geneology and shaping distinctive features. At the same time, the state-list for protection of famous historical and cultural towns, villages and historical buildings has achieved remarkable results and this world heritage protection and management system has been primarily established. In the practice of protection and inheritance of historical and cultural heritage, the protecction and utilization of cultural heritage and cultural innovation-based initiatives have become an important starting point for both demonstrating the cultural content and enhancing the charm of cities.

5.1.1 Overall Protection and Passing Down Civilization

China's cultural traditions and historical civilization have a long history and only its overall protection can ensure that historical imprints do not remain fragmented, and we can protect heritage and pass down civilization more profoundly and comprehensively. Up to now, the State Council has announced 134 national famous historical and cultural cities, 875 historical and cultural blocks have been designated nationwide, and 24,700 historical buildings have been identified. The Ministry of Housing and Urban-Rural Development and the National Cultural Heritage Administration have announced 312 Chinese historical and cultural towns and 487 Chinese historical and cultural villages. Today, the famous historical and cultural cities and towns have become the most comprehensive and systematic vehicle for historical and cultural heritage protection. The delineation of historical and cultural blocks has become an important starting point for expanding the overall protection scope and enhancing the efforts in protection. Historical buildings have become important objects of creative transformation and development of traditional culture.

The *Beijing Urban Master Plan (2016-2035)* endorsed by the Central Committee of the Communist Party of China and the State Council proposed to build a complete system for protection of Chinese famous towns & villages in history & culture with a comprehensive coverage, as well as the overall protection concept of the old town that has been presented throughout time. The purpose of all these efforts is to hand over the historical and cultural heritage of Beijing to the next generation in a complete form.

Since 1982 when Beijing was announced as one of the first famous Chinese historical and cultural cities, its concept of overall protection has been continuously improved upon. In the *Beijing Urban Master Plan (2016-2035)*, it was proposed to continuously explore historical and cultural connotations with a broader perspective, longer-term considerations and more stringent measures, expand the scope of protection, and build the system for protection of famous Chinese historical and cultural cities consisting of four levels, two key regions, three cultural belts, and nine aspects. It also promoted the scope of protection to achieve full coverage of time and space, a more-clearly defined protection structure, a more comprehensive protection content and more abundant protection connotations. Among them, the four levels refer to the protection of famous historical and cultural cities on the four spatial levels of the old town, central city, city area and Beijing-Tianjin-Hebei region. The two key areas refer to strengthening the overall protection of the two key regions of the old town and three mountain regions and five parks. The three culture belts refer to the promotion of the protection and utilization of the of the Grand Canal Culture Belt, the Great Wall Culture Belt, and the Xishan Mountain and Yongding River Culture Belt. The nine aspects refer to strengthening the protection, transmission, and rational utilization of world heritage and cultural relics, historical buildings and industrial heritage sites, historical and cultural blocks and specialty areas, famous towns and villages and traditional villages, scenic and historic areas, historical rivers, lakes and water systems and water cultural heritage sites, landscapes and ancient city ruins, ancient and famous trees as well as intangible cultural heritage.

The old town has always been the core and focus of the protection initiative for Beijing as a famous historical and cultural city. In accordance with the requirements of the Party Central Committee and the State Council on the approval of the *Beijing Urban Master Plan*, "the old town can no longer be demolished, and efforts shall be made to protect the areas that should be protected in the old town through mandated vacating of premises and restorative construction", the old town of Beijing has continuously expanded the targets of protection, increased the scope of the historical and cultural blocks, and included historical buildings, traditional hutongs and historical streets in the protection

Figure 5-1 Structural Planning Diagram of the Protection of Historical and Cultural City in the City Area
(Source: Beijing Municipal Institute of City Planning & Design)

Box5-1 Delineation of Beijing's Historical and Cultural Blocks

From 1993 to 2005, the Beijing Municipal Government has announced 43 historical and cultural blocks in three batches, including 33 blocks in the old town, with a total area of about 20.8km^2, accounting for 33% of the total area of the old town, as well as 10 blocks outside the old town. Three blocks of the Imperial City, Dashilan'r, and Dongsi are listed in the first-batch list of Chinese historical and cultural blocks.

In 2018, in order to implement the protection principle of "protecting the areas that should be protected in the old town" proposed in the new edition of the *Beijing Urban Master Plan*, Beijing sorted out the vast stretches of hutong areas not yet included in the scope of protection in a comprehensive way, and evaluated the protection values and preservation status of each one; at the same time, Beijing carried out protection status assessments of the published historical and cultural blocks, and combined the determination of the protection lists for historical buildings and traditional hutongs in order to formulate the work plan, which would further expand the protected area of historical and cultural blocks.

In the near future, it is expected that Beijing will expand the original protection scope of Dashilan'r and Dongsi South historical and cultural blocks, and incorporate various valuable historical remains in neighboring areas into the blocks in order to implement overall protection. They will also add six historical and cultural blocks into the total incremental area of nearly 4 square kilometers, including Xinjiekou West and Xuanwumen West. In the future, more remaining stretches of hutong areas that meet the delineation standards will be delineated as historical and cultural blocks for protection, and efforts shall be made to gradually strengthen the authentic traditional style of the old town, and to promote the old town's overall protection.

Bird's-eye View of Imperial City Historical and Cultural Block
(Source: Visual China)

Bird's-eye View of Fuchengmennei Historical and Cultural Block
(Source: Visual China)

Guozijian Historical and Cultural Block - Guozijian Street
(Source: Visual China)

system based on strengthening of the protection of the two axes and four layers of enceintes, the historical water systems and the chessboard spatial pattern of Beijing's road network. Relying on the protection of historical and cultural heritage, we will promote the construction of a public cultural service system in the old town, enhance the quality of the heritage area's landscape and environment, and promote the old town's revival.

In December 2017, the Ministry of Housing and Urban-Rural Development listed Beijing, Guangzhou, Suzhou, Yangzhou, Yantai, Hangzhou, Ningbo, Fuzhou, Xiamen and Huangshan as the first batch of 10 pilot cities for protection and utilization of historical buildings in China. This was aimed to strengthen the protection and transmission of cultural heritage and promote the creative transformation and creative development of excellent Chinese traditional culture. It has achieved a series of remarkable results.

Various cities have carried out extensive census and registration of potential historical buildings and established a comprehensive, systematic and complete historical building archive. For example, Huangshan City has conducted censuses on a total of 7438 potential historical buildings, including residential buildings, production buildings, modern public buildings, industrial buildings, etc. of different historical periods and different styles, and after the expert assessments, has determined the final list of 4,845 historical buildings, which was officially announced publicly by the municipal and county governments. Ningbo city also conducted consensus and registered more than 6,500

Figure 5-2　Ancient building of Xidi Town, Huangshan City (Photo by Zhang Guobiao)

Figure 5-3 The Third Batch of Historical Buildings in Ningbo - Baoguo Temple
(Source: http://www.huitu.com)

resource points of historical building, and published a new list of 402 historical buildings on the basis of the published list of 880 historical buildings, increasing the total number of historical buildings to 1,282. Furthermore, Hangzhou city has added a total of 4,054 historical buildings to the reserve list library of historical buildings, and announced a new list of 355 historical buildings, with the total number of historical buildings waiting for official declaration reaching 981.

In addition, various cities have carried out extensive explorations in improving regulations and technical standards, building a region-wide planning framework, exploring innovative utilization models, and establishing supporting policy mechanisms, which have rescued a number of outstanding traditional and modern buildings and thus has promoted a series of protection and utilization practices.

5.1.2 Upholding Ancient and Modern Glory and the Continuation of Context

The preservation of historical and cultural relics not only means the preservation of their originality and authenticity, but also means integrating them into modern life and allowing them to continue playing a role in urban development and residents' lives. For the cities concerned, keeping these memories is not only for the inheritance of culture or the continuation of history, it is also for a better future. The cultural awakening and revival in urban renewal has constantly promoted the integration and development of tradition and modernity, and it is an important force for the passing down of inheritance and urban civilization's roots.

Cultural preservation in the urban built-up area

often requires patience and precision that is needed for "doing embroidery", which means preserving the cultural characteristics and life scenes of specific regions through micro-renewal, and creating a unique temperament that makes the citizens feel intimately connected to the process. Yongqingfang of Guangzhou city is located in the old town, and its underdeveloped infrastructure and uneven building quality make it no longer suitable for modern development and residential needs. But Yongqingfang has Guangzhou's most complete traditional arcade street and distinctive Xiguan architectural culture, as well as the social ecology of the indigenous residents, business and tourists existing in perfect harmony, which therefore contains the city's precious cultural memories. To this end, the local government, with the deep participation of local residents, media, scholars and cultural organizations, used micro-update methods to preserve the spatial texture and external contours of the old buildings. They only updated or repaired necessary components, used modern building materials within the buildings, and improved the living facilities in the neighborhood, so that the region could better adapt to the modern needs of residents and promote the perfect integration of emerging businesses and lives of the original residents while simultaneously retaining the original and authentic old Guangzhou styles and enabling the continuation of collective cultural memories of the old Xiguan. It is widely loved by the citizens of Guangzhou. To cite another example, a lot of alleyways within the ancient city area of Quanzhou also faced the threat of decline due to traditional rapid urban development. In order to rejuvenate the alleyways, Quanzhou city took the micro-renewal of Jinyu Alley as the starting point and conducted exploration. The renewal work adhered to the principle of "micro-interference and low impact",

and carried out improvements such as façade elevation, underground pipelines, ground pavement, night scene optimization, street greening, street furniture layout, etc. without changing the lifestyle of the indigenous people. These improvements activated the old streets and towns, improved the living conditions of the indigenous people and preserved the space fragments of different historical stages. Most of the people involved in this project were local professionals of the ancient city who were familiar with the history, culture and ancient architecture of the ancient city of Quanzhou. Most of the craftsmen were over 50 years old with rich experience and their work reflects the traditional craftsmanship of ancient buildings in south Fujian, retains the memories of the people, reflects their longing for home, and finally, attracts many tourists to experience the Minnan culture of south Fujian province.

Continuation of the cultural context requires both attention to the people and the full participation of people from all walks of life in the process of governance updates in order to promote the revival of the old town in a more moderate and sustainable way. As the oldest port in Xiamen, Shapowei is located in the old town of Xiamen and boasts many shipyards and docks, a dense population living and working there, thriving and prosperous businesses, as well as abundant resources such as Danmin culture, fishing culture, handcraftsmanship culture and traditional community culture. However, with the outflow of the population in the old town, the aging infrastructure and disorderly growth of unauthorized construction, traditional culture is on the verge of extinction, regional vitality is rapidly declining, and the old town became a deserted place in Xiamen. In the subsequent rounds of transformation, due to the failure to pay attention to and focus on regional cultural continuity and the social dilemmas and difficulties of livelihood, the transformation had not been recognized by relevant interest groups and therefore it was difficult to advance the transformation. In 2016, the role of the government underwent major adjustments from that of policy maker and leader to that of promoter and participant, and successfully established the "Shapowei Co-construction Workshop" with the professionals of the universities as the main body. By taking a workshop as a platform on which the various fields of society could engage in thinking and discussions, the government has promoted aboriginal people and social forces with humanistic sentiments to actively participate in the process of regional renewal. There are

Figure 5-4 Real Scene Picture of Yongqingfang, Guangzhou (Left)
(Source: http://www.huitu.com)
Figure 5-5 Real Scene Picture of Jinyuxiang, Quanzhou (Right)
(Source: http://www.huitu.com)

Figure 5-6 Real Scene Picture of Shapowei, Xiamen
(Source: http://www.quanjing.com)

doing so by cultivating the cultural groups dominated by fishermen, sorting out the history of Shapowei and folk oral history, excavating old objects, conducting public consultation meetings, and establishing an effective "community service learning workshop" mechanism, etc. At the same time, the local government has strengthened supervision and effective restraint on the administrative power and capital holders, explored the formation of an urban spatial governance structure that is jointly operated by the government, the market, and social forces, and promoted consensus among all stakeholders to promote the renewal process and sustainable development. This has not only improved the living conditions of the aboriginal people to the maximum degree possible, but also allowed them to share the dividends of the district renewal, thus fully extending the social ecological relationship and historical and cultural context of Shapowei.

5.2 People-oriented Public Space

In 2017, the Ministry of Housing and Urban-Rural Development announced the first and second batch of 57 urban design pilot cities. These cities have made much progress in setting up departmental institutions, establishing institutional norms, carrying out application practices, and promoting relevant governmental organs to carry out refined design and management of various public spaces in the cities. The urban design system has continuously improved, the field has been continuously expanded, and the concepts have been continuously updated. Great achievements have been made especially in the comprehensive management of street space, the open regeneration of waterfront space, the dynamic shaping of new types of spaces and the activation and utilization of special spaces.

5.2.1 Comprehensive Management of Street Space

Streets are the space with the highest concentration of various spatial elements and the place for the most frequent urban public activities in the cities. In the development process of the past few decades, street construction and management have been mainly focused on improving the organization efficiency of motor vehicle transportation. The insufficient understanding of the complexity of street functions and inadequate attention to the needs of the people have led to chaotic development of the street space, making it impossible for the streets to meet the growing needs of peoples' public life.

In recent years, the streets have received unprecedented attention in the field of urban design and have become an important target for urban renewal and governance. Based on the transformation from vehicle-first to people-first, relevant guidelines have been introduced in more than ten cities and regions, including Shanghai, Nanjing, Guangzhou, Luohu District of Shenzhen, Beijing, Foshan, Zhuzhou, Xiamen, Hangzhou and Kunming. The *Shanghai Street Design Guidelines* explores the transformation of street design concepts for the first time and advocates consensus among all sectors of society. The *Guangzhou Complete Street Design Manual* integrates all-factor planning and design standards to guide the integration of both functional and landscape design into the road construction and transformation process to realize the transformation of traffic-oriented roads to life-oriented roads; the *Beijing Urban Design Guidelines for Street Renewal and Governance* focuses on urban renewal and governance from the perspective of innovative planning and coordination; the *Nanjing Street Design Guidelines* focuses on proposing the implementation strategy of street design, promoting human-oriented transformation of the streets and creating a city image with Nanjing characteristics.

At the same time, street renovation and transformation work has been carried out extensively, and many practical cases with benchmarking significance have emerged. As an important traffic trunk line in the waterfront area, Yangshupu Road of Shanghai has taken historical protection and landscape upgrading as its premise in its comprehensive renovation and renewal process, abandoned the usual practice of road reconstruction, and focused on the overall protection of buildings on both sides of the road. Furthermore, the transformation has coordinated a consideration of harmony between traffic and urban styles, created a refined design of the street section instead of adopting standard sections, and kept most of the existing platanus orientalis by reducing the width of the motorway to 3.25 meters, thus fully retaining urban memories, spatial patterns, street scales and historical relics of the area. It not only solved the problems of travel and livelihood of 150,000 residents along the road, but also promoted the mutual promotion of historical relics and regional vitality, as well as the organic integration of street space and public activities.

5.2.2 Open Regeneration Waterfront Space

Today, with the increasing emphasis on water environments, the urban waterfront space environment has become a hot issue in urban design. The renewal of waterfront space has become an important way to improve urban landscapes and the human settlement environments and also to promote urban economic development. In recent years, the waterfront space remediation in many cities of China has focused not only on water management and ecological restoration, but also on the opening and revival of waterfront space.

The focus of waterfront space remediation first of all lies in forming a continuously straight, open and accessible space, which means reorganizing the waterfront space and giving the waterfront space proper functions and a pleasant quality. More importantly, it means connecting more cultural resources through linear space, and promoting their retention, exhibition and reutilization. Huangpu River Open Space Regeneration Project in Shanghai opened up 45 kilometers-long of open space along the river for the public by coordinating with organizations such as the military troops, central enterprises, customs border inspection authorities and state-owned enterprises to vacate the spaces along the

Figure 5-7　Comparison of the Actual Scene and Planned Design of Yangshupu Road from the Same Perspective
(Source: Shanghai Urban Planning and Natural Resources Bureau)

river banks (some organizations were required to vacate before negotiation) and encourage construction units to invest in demonstration projects. They created open jogging trails, promenades, bicycle lanes, and more and more hydrophilic spaces such as urban living rooms for the citizens, trying their best to preserve the historical features and historical relics along the banks while also implanting new urban functions and cultivating new urban culture. Through public space remediation and improvement, the plan for the slow path system on both sides of the Pearl River in Guangzhou formed. They created a continuous 60-kilometer-long Pearl River slow path system-promenades plus jogging trails, plus the riverside bicycle lanes; at the same time, by increasing the width of the public space belt, they enhanced the inter-visibility and accessibility of the corridors on both sides of the river, comprehensively optimizing the quality of the landscape along the river and promoting the protection, restoration and activation of historical and cultural blocks as well as historical buildings and industrial heritage. As a result, this not only perpetuates the characteristic life scenes of Guangzhou, but also provides innovative power for development along the Pearl River. On the basis of the Zuohai-Xihu Lake connection project and the river improvement projects of the Luzhuang River, Chating River, Yangli River, Niupu River, Longjin River

Figure 5-8 The waterfront area of Shanghai North Bund offers new scenic perspectives for tourists to Huangpu River (Photo by Hai Niu)
(Source: http://www.cnsphoto.com)

Figure 5-9 Real Scene Picture of Riverside Space of Pearl River in Guangzhou
(Source: http://www.huitu.com)

and the upstream section of Jingang River, Fuzhou has continued to promote the renovation of 61 river courses of the city, and has constructed the riverside greenway with the total length of 400 kilometers to form a beaded park green space on both banks of the inland rivers. At the same time, Fuzhou has paid attention to preserving ancient trees and monuments and creating an ecological revetment in the design of the beaded park, provided the facilities that allow tourists to walk, stop or stay to have fun by offering tree-lined squares, waterside space and service facilities, and improved the residents' experience by installing fitness equipment and 24-hour library facilities.

5.2.3 Space-building for Specific Population Groups

People's requirements for public space are constantly diversified, and urban planning and construction is also trying to explore the types of spaces that meet the needs of various groups of people. In recent years, new spaces have been mainly concentrated in sports and leisure spaces as well as child-friendly spaces.

With the germination of people's health and leisure consciousness and the pursuit of a leisurely life, leisure in the form of sports has gradually become one of the mainstream trends of citizens today. The *Outline of the Healthy China 2030 Plan* issued by the CPC Central Committee and the State Council on October 25, 2016 proposes to "coordinate the construction of the public facilities for public health, and strengthen the construction of site facilities such as fitness trails, cycling lanes, national fitness centers, sports parks, and community multi-purpose sports fields." Against this backdrop, many cities use sports elements to stimulate urban vitality, and increase the construction of sports and leisure spaces in urban renewal. Pujiang County of Jinhua City has built ecological corridors on both sides of the 210 Provincial Highway connecting the east and west of Pujiang County, designed 18 kilometers of bicycle lanes, 12 kilometers of trails and 6.5 kilometers of plank roads, and set up 5 sports stadiums in the corridors, which has become the unique dynamic landscape of Pujiang. Taihu Lake Bay "Sports and Leisure Town" in Changzhou City integrates such elements as the shaping of sports leisure space, supporting services for sports facilities and the construction of sports training bases into tourism planning and design, and allows people to engage in sports activities such as dragon boat racing, skiing, fishing and cycling. At the same time, a large number of open semi-professional sports and leisure venues have emerged in Beijing, Shanghai, Chengdu, Hangzhou and other cities, and these venues are constructed by the communities or commercial entities or sports venues and serve as public institutional facilities but are open to the public. These venues have greatly satisfied the healthy living needs of the public in terms of space. In addition, governments of many cities such as Gaoyou City, Chenzhou City, Luoyang City, Yangzhou City, and Nanning City have invested in building sports and leisure parks to guide residents to participate in sports during leisure

Figure 5-10 100,000 walkers walk in Tianjin Sports Theme Park (Photo by Zhang Daozheng)
(Source: http://www.cnsphoto.com)

and recreation time, for bodybuilding and stimulating the dynamic or inspiring the urban, public atmosphere.

At the same time, children and the elderly are increasingly valued by urban builders, who have incorporated the fundamental demands of children in the planning of blocks or cities in particular. Since 2001, the Shenzhen Municipal Government has issued and implemented three rounds of children's development plans, and has incorporated "actively promoting the construction of a child-friendly city" into the 13th Five-Year Plan for Economic and Social Development in the Shenzhen Municipality. The *Shenzhen Child-Friendly City Action Plan* (2018-2020) released on February 6, 2018 focused on the promotion of children's well-being with four aspects: ensuring safety, expanding space, encouraging participation, and enhancing protection, and identified six tasks of: promoting children's safety and security, expanding child-friendly space, enhancing children's participation in practice, improving children's social security, popularizing and promoting the concept of child-friendly concepts, and studying children's public policies. Moreover, Shenzhen also selected commu-

Figure 5-11 Shenzhen launches Children's Friendly City Planning and Development (Photo by Ren Yongdong & Liu Lei)
(Source: Urban Planning & Design Institute of Shenzhen)

nities, schools, libraries, hospitals and other fields closely related to children's growth and most-visited by children as the pilot venues to carry out child-friendly construction projects, and as the locales in which to formulate guidelines for child-friendly construction in related fields. These were then promoted throughout the city. Changsha City has also incorporated the construction of a child-friendly society into the *Changsha 2050 Vision Strategic Development Plan* being prepared, and the city has carried out the preparation of the *Child-Friendly City Planning Guidelines Study* and formulated guiding standards and construction guidelines for a child-friendly city in Changsha. At the same time, Changsha selected 10 primary schools in the city and proposed an independent, continuous and safe walking path plan for the complex surrounding environment outside the campus in order to eliminate potential safety hazards and alleviate traffic congestion during the peak period of going to and from schools. In addition, children were involved in the planning and design in various forms, and children were organized to paint 290 com-passionate zebra crossing signs, pedestrian crossings at bus stops, traffic signs, municipal manhole covers, etc., and their excellent works were used in the implementation plan.

5.2.4 Activation and Utilization of Vacant Spaces

Establishing a design and management platform to bring together the wisdom and strength of all sectors of society is becoming an important means for cities to collaborate and progressively carry out urban public space renewal and activation. In this process and based on extensive surveys and research, major cities continue to explore special types of space, focus on urban gray spaces such as corner spaces, under-bridge spaces, and abandoned spaces prompting their activation and regeneration, and to provide more urban open spaces with vitality and quality to the cities concerned.

Shanghai, Beijing, Shenzhen and other cities have promoted planning designers to enter the communities by organizing design competitions establishing multi-party participation platforms. This has contributed to many examples of small- and micro-space transformations that are popular among the public. The cross-over design competition platform, entitled Shenzhen Small & Smart Design Urban Micro-Design, and initiated by Shenzhen Center for Design, made full use of the institutional advantages of the residents' council and conducted targeted improvement design and renovation implementation for various types of small and micro spaces based on the needs of the public in communities. Thus, they have established a good long-term relationship between the communities and the planning designers.

Figure 5-12 The First "I Love You" Zebra Crossing in Jiujiang, Jiangxi Province (Photo by Ouyang Haiyuan) (Source: http://www.cnsphoto.com)

Figure 5-13　Micro Remediation Results of Longling Community Playground in the Shenzhen Small & Smart Design Competition
(Source: Shenzhen Center for Design)

Figure 5-14　Xiamen Railway Park
(Source: http://www.huitu.com)

Figure 5-15　Railway Park, Wuluju, Haidian District, Beijing
(Source: Photo by Author)

In this process, the idle space under the viaducts entered the minds of the governments and the public. Considering the shortage of land resources in urban centers on one hand, and the ever-increasing high public demand for public activities on the other, the utilization of space under the viaducts has become a focus of social discussion and practice. Guangzhou has transformed the space under the viaduct of the ring road near its Olympic Center into sports fields such as football fields, basketball courts, badminton courts, ice rinks, etc., providing cost-effective and adequate sports facilities for surrounding residents and sports enthusiasts. Wenzhou has also transformed the space under the viaducts of the highways into sports fields, but also cultural parks and a people's stage and rest spaces, which has brought a space with a diverse sense of vitality to the surrounding communities.

With the iteration of urban industrial development, the railways that once boosted industrial development gradually withdrew from the historical stage, and abandoned railways have become an important target for shaping public spaces with urban characteristics. Tianjin and Xiamen have upgraded the space along the abandoned railways and opened them as urban linear parks, which are now known as Tianjin Greenway Park and Xiamen Railway Park respectively. On the basis of improving the quality of public spaces along the railways, Guangzhou has also implanted certain facilities and functions in the surrounding industrial plants along the railways to stimulate the cultural value of industrial heritage, promote the transformation and renewal of the surrounding areas, and enhance the vitality of the city. Nowadays, the space renewal path of the abandoned railways has gradually matured and become an important entry point for the transformation of resource-based cities.

5.3 China Program: Industrial Heritage Protection and Reuse

Upon entry into the 21st century, China initiated a historic transformation from "the protection of cultural relics" to "the protection of cultural heritages". In 2005, the State Council issued the *Notice of the State Council on Strengthening Protection of Cultural Heritage*, which was the first time that the concept of "cultural relics" has been replaced by the concept of "cultural heritage" in national-level official documents, and the first time that industrial heritage has entered the discussion of protection.

Looking back on the reform and opening up, China's industrial development has created great material wealth and produced brilliant ideological progress and wealth at the same time. With the change of economic and social development and industrial models, traditional industry has experienced a process from prosperity to depression to transformation and upgrading. China's industrial heritage protection work has also experienced the process of germination, discussion, exploration and vigorous development and has become the focus of urban renewal construction in recent years.

5.3.1 National Measures and Lists Promulgated: China Industrial Heritage Protection List (First Batch & Second Batch)

On January 27, 2018, National Academy of Innovation Strategy and Urban Planning Society of China officially announced the China Industrial Heritage Protection List (First Batch), and it covered 100 projects including Jiangnan Machinery Manufacturing Bureau, Hanyang Iron Works and Beijing-Zhangjiakou Railway

Box5-2　The Development History of China's Industrial Heritage Protection in the 40 years of Reform and Opening Up

- Since the reform and opening up, under the demand of China's economic transformation and development, traditional industries have gradually withdrawn from the historical stage in the cities with relatively advanced economic development. Many industrial plants, machinery, equipment and facilities are no longer of production significance, and are abandoned or dismantled and this has triggered people's deep feelings about the past regarding work and life.
- By the end of the 20th century, China's cultural industry had developed rapidly. The abandoned industrial plants and parks have become the new spaces of cultural and creative industries because of their low rental prices, convenient geographical locations and industrial styles. Many exploratory transformations of industrial buildings and industrial plants have been fruitful, such as Beijing's 798 art district, Shanghai's Tianzifang arts and crafts enclave, and Guangzhou's Zhongshan Shipyard Park, which have created a large international impact and triggered a heated discussion among the public about the "dismantling and not dismantling" of abandoned factories.
- In April 2006, the China Industrial Heritage Protection Forum was held in Wuxi and adopted the, *Wuxi Initiative-Attaching Importance to the Protection of Industrial Heritage during High-speed Economic Development*, which was officially promulgated by the National Cultural Heritage Administration in June 2006 and became the first constitutional document on industrial heritage protection and research in China. It was also an important starting point for the research of industrial heritage in China. The concept of "industrial heritage" has officially entered China's vision. Local policies have been continuously released, and relevant academic research and practical explorations have shown an exponential growth.
- In the third national cultural relics survey conducted by the State Council from 2007 to 2011, "industrial heritage" was raised as a special topic, which greatly mobilized the enthusiasm of all provinces and cities to excavate industrial heritage, thus the degree of recognition for industrial heritage has gradually improved.
- From 2010 to 2015, the relevant organizations and institutions for the research and protection of industrial heritage under the National Cultural Heritage Administration, the Ministry of Housing and Urban-Rural Development, the Ministry of Industry and Information Technology, the People's Political Consultative Conference, and China Association for Science and Technology were gradually established, and they advanced the transformation and upgrading of traditional old industrial bases and promoted the development of industry, informatization and their convergent development to become a new path for the protection and reutilization of industrial heritage sites.
- With the gradual clarity of the concept of industrial heritage and the establishment of the organizations for its protection, research and management of industrial heritage, industrial heritage sites have highlighted their own unique value categories and cultural significances. On January 27, 2018, China Association for Science and Technology and the Urban Planning Society of China officially released the China Industrial Heritage Protection List (First Batch), and a total of 100 projects were listed covering shipbuilding, military industries, railway and other categories. On April 12, 2019, the China Industrial Heritage Protection List (Second Batch) was officially released, and included 100 industrial heritage sites.

(See Appendix 1-1 for details). This list includes the government-run enterprises that were created during the Westernization Movement and also the key construction projects of the "156 Soviet Assisted Projects" after the founding of New China. They cover a wide range of shipbuilding, the military industry, and railway sectors. The are the representative industrial heritage sites with outstanding value. On April 12, 2019, the China Industrial Heritage Protection List (second batch) was officially released, and covered 100 industrial heritage sites (see Appendix 1-2 for details). This list covers important heritage projects during the period from the Westernization Movement to the National Third-Front Movement, covering mining, metallurgy, the military industry, transportation, machinery, light industry and other industries.

The issuance of two China industrial heritage protection lists has clarified 200 heritage projects, continuously expanding the connotations of industrial heritage, and increased the time span of industrial heritage recognition, which not only helps to preserve a complete and rich imprint from the industrial civilization, but also lays a solid foundation for the discovery, protection and utilization of the values of industrial heritage sites in the new era.

In this critical period of transformation of urban development patterns and economic development upgrading, the value of industrial heritage protection has become increasingly prominent. For a long time, industry has been the core element of urban development, and industrial heritage has become an essential element of urban culture. How to achieve this essential inheritance is a new topic that needs to be explored. In the process of activation and utilization of heritage, holistic and systematic thinking is very important. We should focus on special urban features, pay attention to industrial development and functional optimization, and integrate industrial heritage into contemporary life, making it an important part of public life.

5.3.2 Local Practice: Overall Protection and Renewal of Beijing Shougang Industrial Heritage Sites

The predecessor of Shougang Old Industrial Base was the Longyan Iron Mine Limited, founded in 1919. It has a history of 100 years and is the cradle of the steel industry of the Republic and the banner of reform and opening up of the P.R. China. At the end of the last century, with the advent of the post-industrial era and the intensification of capital pollution control, Shougang's development encountered unprecedented challenges. In 2005, in order to bring the blue sky to the capital city, the steel production function of Shougang Old Industrial Zone was moved to Caofeidian of Tangshan city; and in 2010, Shougang completely stopped production and began to explore the path of comprehensive transformation and development.

The large-scale old steel industry factory area is both a difficult problem and a precious legacy with considerable value. The protection and renewal of Shougang Industrial Heritage has identified a multi-purpose transformation development strategy such as environmental improvement, cultural protection, and economic and social synergy. In order to change the basic conditions, Shougang has adhered to reducing the production and increasing green space, restoring the industrial land to green spaces, and has planned to build a 1.8-square-kilometer Yongding River Riverside landscape belt to showcase the characteristic landscape of the urban mountain-water-industrial heritage convergence. On this basis, Shougang has combined industrial transformation with urban innovation and development, built a sponge system capable of surviving heavy rains (once every 50 years) with zero impact with the basic concept of green ecology, achieving a recycling rate of 90% for construction waste and 100% green buildings. It has launched projects of intelligent parking system construction and operation, and pollution and soil environment control, which became China's first and the world's 19th project to be part of the C40 Climate Positive Development Program in 2016. These efforts enabled Shougang to successfully transform itself from a large steel company into a comprehensive urban service provider. In the process, nearly 10,000 employees who were left behind were then re-employed, ensuring the stable social and economic development during this transition period.

The forthcoming 2022 Beijing Winter Olympics will inject more vitality into Shougang's development. In 2015, the Winter Olympics Organizing Committee settled in Shougang, revitalized the industrial resources into ice and snow sports leisure facilities, and built the National Winter Sports Training Center. They have undertaken high-end "sports+" industries such as sports technology, sports media and sports creative industries, and remained committed to building the national sports industry demonstration zone. At the same time, under the joint promotion of the planning and design team with Shougang as the implementation entity, the silo creative space, blast furnace museum, industrial heritage park, desulfurization workshop intelligent center and other projects were implemented, which preserved and utilized the industrial relics in creative ways and inspired the vitality of the post-industrial era sites.

5.3.3 Local Practice: Comprehensive Protection and Development of Modern Ceramic Industrial Heritage in Jingdezhen

Jingdezhen is an ancient city that has prospered from the traditional ceramics industry. It is also a modern ceramic industrial city with a reputation both at home and abroad. There used to be porcelain kilns and factories everywhere in Jingdezhen. During the mid-to-

Figure 5-16 Real Scene Picture of New Shougang
(Source: Photo by Author)

Figure 5-17 Real Scene Picture of the New Shougang Ice and Snow Sports Facilities (Photo by Fu Tian)
(Source: http://www.cnsphoto.com)

late 1990s, a large number of porcelain kilns and factories were forced to shut down due to the decline of the old industry and, therefore, Jingdezhen withdrew from the historical stage. For a long time, a large number of old factories in the old town remained abandoned, resulting in many "black holes" in the city's capacity.

In March 2017, Jingdezhen was listed as one of the second batch of pilot cities for ecological restoration and city betterment in China. The local government fully promoted the integration of ecological restoration, urban betterment and urban multi-innovation development, while focusing on organic integration of the protection of famous national historical and cultural cities as well as the high-quality development goals of the city. They incorporated the protection, renewal and renovation of the old factory area into the overall urban development plan, and rectified the environment of 14 old factories in Jingdezhen to create a repositioning of the industry. The transformation of the old factory area adheres to the principles of retaining the status quo, restoring the original form and repairing the old as much as possible, and fully retains the development culture of the ceramics industry. After the restoration, the Xujia Kiln became an important place for non-tangible ceramic cultural heritage and an indispensable part of the heritage network of Jingdezhen ceramics industry; through the improvement of the existing buildings in terms of comfort, the renovated Yuzhou Ceramics Factory was transformed into a museum and an art gallery, retaining the original feature of the kiln; during the renovation, other surrounding plants have become new spaces related to catering and art exhibitions, realizing a fusion of historical places with modern times.

In the process of transformation, the factories have always adhered to the principles of closely combining historical culture, urban status and community

Figure 5-18　Taoxichuan Industrial Heritage Museum
(Source: http://www.huitu.com)

Figure 5-19　Jingdezhen Heritage of Ceramic Industry Museum (Photo by Zhong Xin)
(Source: http://www.cnsphoto.com)

environment. They have greatly stimulated the industrial innovation development of Jingdezhen and revitalized the urban economy through various means such as passing on skills, retaining community memories, and the modernization of transformation. More importantly, through transformation, the factory area has become a young and open international community, and provided local citizens with a number of high-quality places for leisure activities.

5.3.4 Local Practice: Reopening and Utilization of Columbia Circle, Shanghai

One hundred years ago, Columbia Circle was once a place for foreign expatriates to entertain themselves. After the founding of the People's Republic of China, the Shanghai Institute of Biological Products moved its office to this place and it remained closed to the public for nearly 70 years. The park consists of three historical conservation buildings, 11 industrial buildings that have been standing throughout the history of the People's Republic of China, and four distinctive contemporary buildings. These old buildings, along with the surrounding old-style garden houses of Panyu Road and Xinhua Road, represent the city identity of Shanghai in the memories of old Shanghaiese.

The spatial pattern in the compound has been preserved to the utmost extent, the buildings have been restored to their original taste, and even the historical interior styles have been restored. For example,

Figure 5-20   Real Scene Picture of Public Space, Columbia Circle
(Source: http://www.huitu.com)

the former Columbia Country Club retained the gym and swimming pool of its time. The MMR vaccine production building also retained the unique Bauhaus architectural style, with the windows partially transformed to accommodate new functional requirements. The high-level architectural art works that survived these different historical periods exhibit the memories of life and work in all these periods of time. After the new functions were implanted, the buildings share a symbiosis with the diverse lifestyles they accommodate. While passing on the historical context and improving the efficiency of the land, the heritage renovation project also brings about 12,500 square meters of public space to the local citizens and 7,000 square meters of public service facilities, which is particularly valuable in the central area of Shanghai, and provides a brand new way of thinking about transformation of industrial heritage in an urban downtown area.

Figure 5-21　Former Mansion of Sun Ke, the Son of Dr. Sun Yat-sen, in Columbia Circle, Shanghai
(Source: http://www.huitu.com)

Chapter 6

Rural Revitalization and Poverty Alleviation

Rural Revitalization Strategy

Poverty Alleviation

Improving Rural Living Environment

Small Cities and Towns Construction and Characteristic Development

China's Rural Practice

Rural Revitalization and Poverty Alleviation

China's rapid urbanization process has played an important role in social and economic development, but some problems have accumulated. In response to new problems and challenges, the 18th National Congress of the Communist Party of China proposed a new-type urbanization strategy in 2012, emphasizing a shift of the development mode from focusing on quantity to focusing on quality. The social and economic development of rural areas in China is the basis of urbanization. However, some remaining issues have exposed the weak links of modernization such as needing to improve the quality of agricultural supply, the insufficient ability of farmers to adapt to productivity development and market competition, the prominent issues in the rural ecological environment, and the incomplete mechanism for reasonable flow of factors between urban and rural areas. In 2017, the 19th National Congress of the Communist Party of China proposed a strategy for rural revitalization, giving priority to promoting rural social and economic development by mobilizing resources throughout China. The development of China in the new era will feature the mutual reinforcement and double-track approach using these two strategies. New-type urbanization is the only way towards modernization and an important way to solve the problems facing agriculture, farmers and rural areas. The rural revitalization strategy opened up a new situation necessitating urban-rural integration and modernization, and is the cornerstone for promoting new-type urbanization.

6.1 Rural Revitalization Strategy

6.1.1 Rural Revitalization Strategy and Implementation Path

In October 2017, the 19th National Congress of the Communist Party of China put forward the strategy for rural revitalization, which proposed adhering to the priority development of agriculture and rural areas, and establishing and improving the institutional mechanism and policy system for the integrated development of urban and rural areas. Furthermore, it aims to accelerate the modernization of agricultural and rural areas in accordance with the general requirements of "building rural areas with thriving businesses, pleasant living environments, social etiquette and civility, effective governance, and prosperity".

In January 2018, the Chinese government issued the *Opinions on Implementing the Rural Revitalization Strategy*, which further explained the background of the rural revitalization strategy, analyzed the current problems in rural development, and divided the implementation of the rural revitalization strategy into three stages: by 2020 rural revitalization will have made important progress, and the overall institutional framework and policy system will have essentially taken shape; by 2035, rural revitalization will have made decisive progress, and agricultural and rural modernization will have fundamentally been realized; by 2050, the following goals will be realized: rural areas will have been fully revitalized, agriculture will be strengthened, rural areas will be beautified, and farmers will achieve prosperity. The core of the rural revitalization strategy includes improving the basic rural management system; deepening the commitment to reforming the rural land system; deepening the commitment to reforming

Main Indicators of Strategy Plan for Rural Revitalization Table 6-1

Classification	No.	Main Indicators	Unit	2016 Base Value	2020 Target Value	2022 Target Value	Increase in 2022 Compared to 2016 [accumulated percentage points]	Attributes
Thriving Businesses	1	Integrated Grain Production Capacity	100 million ton	>6	>6	>6	—	Obligatory
	2	Contribution of Agricultural Science and Technology Progress	%	56.7	60	61.5	[4.8]	Anticipated
	3	Agricultural Labor Productivity	10,000 Yuan/person	3.1	4.7	5.5	2.4	Anticipated
	4	Ratio of Agricultural Product Processing Output Value to Total Agricultural Output Value	—	2.2	2.4	2.5	0.3	Anticipated
	5	Tourists for Leisure Agriculture and Rural tourism	100 million visits	21	28	32	11	Anticipated
Pleasant Living Environments	6	Comprehensive Utilization Rate of Livestock and Poultry Manure	%	60	75	78	[18]	Obligatory
	7	Village Green Coverage Rate	%	20	30	32	[12]	Anticipated

Classification	No.	Main Indicators	Unit	2016 Base Value	2020 Target Value	2022 Target Value	Increase in 2022 Compared to 2016 [accumulated percentage points]	Attributes
Pleasant Living Environments	8	Proportion of Villages That Handle Domestic Waste	%	65	90	>90	[>25]	Anticipated
	9	Penetration Rate of Rural Sanitation Toilets	%	80.3	85	>85	[>4.7]	Anticipated
Social Etiquette and Civility	10	Village Comprehensive Cultural Service Center Coverage Rate	%	—	95	98	—	Anticipated
	11	County-level and Above Civilized Villages and Townships	%	21.2	50	>50	[>28.8]	Anticipated
	12	The Proportion of Full-time Undergraduate Education in Rural Compulsory Education Schools	%	55.9	65	68	[12.1]	Anticipated
	13	Rural Residents' Education, Culture and Entertainment Expenditure	%	10.6	12.6	13.6	[3]	Anticipated
Effective Governance	14	Coverage Rate of Village Planning and Management	%	—	80	90	—	Anticipated
	15	Proportion of Villages With Integrated Service Stations	%	14.3	50	53	[38.7]	Anticipated
	16	Proportion of Villages With Village-level CPC Secretary Working Concurrently as the Village Committee Director	%	30	35	50	[20]	Anticipated
	17	Proportion of Villages with Village Regulations and Folk Conventions	%	98	100	100	[2]	Anticipated
	18	Proportion of Villages with a Strong Collective Economy	%	5.3	8	9	[3.7]	Anticipated
Prosperity	19	Engel'n Coefficient of Rural residents	%	32.2	30.2	29.2	[-3]	Anticipated
	20	Ratio of Urban Income to Rural Income Per Capita	—	2.72	2.69	2.67	-0.05	Anticipated
	21	Penetration Rate of Tap Water in Rural Areas	%	79	83	85	[6]	Anticipated
	22	Proportion of Incorporated Villages With Paved Roads Where Conditions Allow	%	96.7	100	100	[3.3]	Obligatory

Note: The "village" that is not specifically identified in this indicator system and plan refers to the area under the jurisdictions of the village committees and the agriculture-related committees.
(Source: Strategy Plan for Rural Revitalization (2018-2022))

the rural collective property rights system; ensuring national food security; improving the agricultural support and protection system; promoting the integrated development of the primary, secondary and tertiary industries in rural areas; improving the rural governance system; and fostering the work teams to serve agriculture,rural areas, farmers and so on.

The implementation path of the whole strategy will be carried out from various aspects, including reshaping the urban-rural relationship, consolidating and improving the basic rural management system, deepening the structural reform of the supply side of agriculture, adhering to the harmonious coexistence of man and nature, perpetuating and developing agricultural civilization, innovating the rural governance system, carrying out the campaign of targeted poverty alleviation, and focusing on the measures of promoting industrial, talent, cultural, ecological and organizational revitalization in rural areas.

6.1.2 Policy Promotes the Integrated Development of Urban and Rural Areas

In September 2018, the Chinese government proposed in the *Strategy Plan for Rural Revitalization (2018-2022)* to build a new spatial pattern of rural revitalization and improve measures to promote the policy system of integrated development in urban and rural areas. Through this top-level design, China can promote the integrated development of urban and rural areas, thereby narrowing the gap within China and simultaneously move towards environmentally sustainable development goals through structural transformation.

The new pattern of rural revitalization emphasizes the overall development of urban and rural land and space; the optimization of rural production, living spaces and ecological spaces; and the promotion of rural revitalization by different categories. The urban-rural layout structure primarily proposes to focus on constructing the urban pattern of coordinated development of large, medium and small cities and small towns with urban agglomerations as the main body while strengthening the impetus for urban areas to consider rural areas. Secondarily, the structure proposes the acceleration of development in small and medium-sized cities, improves the comprehensive service functions of counties, and promotes the transfer of agricultural populations to local and nearby urbanized areas; On a tertiary level, the structure proposes to develop towns and small cities with unique characteristics according to local conditions, promotes villages and towns towards reciprocal development, and promotes the joint development of towns and villages; Finally, the structure proposes to build beautiful and ecologically livable villages that provide multiple functions and high-quality products, while also passing on rural culture, retaining the memories and nostalgia of our homeland, and satisfying the people's growing needs for a better life.

In addition to material space policies, the urban-rural integration development policies include policies on talent, land, and funds related to rural revitalization, and emphasize the double-track approach to the urban and rural mobility of these three components. First, policies on talent emphasize accelerating the urbanization of rural migrants, and at the same time promoting the revitalization of rural talent by encouraging social talent to join in rural development. Secondly, the reform of the rural land system has continued to advance:on the one hand, it established and improved the homestead management system featuring legal and fair acquisition, thrifty and intensive utilization, and voluntary exit with compensation; and on the other hand, it promoted the market access of collectively-owned commercial construction land, and granted the rights and functions of transfer, lease, and equity of this land. Finally, financial policies take priority in fiscal security as the core to fostering rural construction and development, and at the same time stimulate the motivation and vitality of social investment while allocating financial resources to key areas or weak links in rural economic and social development.

6.1.3 Planning Leads the Rural Revitalization and Development

The rural revitalization strategy has always emphasized the leading role of planning. Promoting unified planning of urban and rural areas is one of the important measures to building a new pattern of rural revitalization and coordinating urban and rural development. The current planning tasks include coordinating and planning the main layout of urban and rural industrial development, infrastructure, public services, resources and energy, and ecological environmental protection. Furthermore, forming an urban and rural development pattern with both rural villages and modern towns possessing their own characteristics; strengthening the guidance and constraining roles of county-wide space planning, scientifically arranging the county-wide rural layout and resource utilization, facility allocation and village rectification, and promoting the full coverage of village planning and management. Tasks also include comprehensively considering the villages' evolution law, agglomeration characteristics and current distribution, and reasonably determining the county-wide village layout and scale by combining farmers' production with living radius; strengthening the overall control of rural styles, paying attention to the individual design of rural houses, building the countryside based on the rural community, rich in regional characteristics, bearing in mind rural and pastoral nostalgia, and reflecting modern civilization while preventing the urbanization of rural landscape.

Since September 2018, the Ministry of Housing and Urban-Rural Development has promoted the "design down-to-village" initiative nationwide, guiding and supporting designers in the fields of planning, architecture, landscape, municipal engineering, art design, and cultural planning to offer services at the village level, solving out standing problems of rural human settlements considering actual situations and village needs, and more importantly, promoting capacity building of rural communities. The "design down-to-village" initiative is a kind of companionship service that fully respects the opinions of the villagers, practices joint planning and decision-making, joint construction and mutual development, joint management of construction, joint evaluation of results, including shared results, and pays attention to cultivating local talents such as those of rural craftsmen. In February 2019, on the basis of this initiative, the services featured by "joint planning, joint construction, joint management, joint evaluation, and sharing" were extended to urban and rural communities, and the campaign of "joint creation of a beautiful environment and a happy life" was carried out in the construction and renovation of urban and rural living environments, which promoted the formation of a coordinated urban and rural governance system while improving living environments.

Box 6-1　The First National "Green Point Contest" Focuses on Rural Revitalization

The Green Point Contest, as one of the four major sections of the Green Development Innovation Conference, was hosted by the Urban Planning Society of China, the Sichuan Provincial Department of Housing and Urban-Rural Development and the Suining Municipal People's Government in 2018 with the theme of "Village Revitalization and Settlement Renewal."The aim was to create a number of excellent rural settlement plans and designs, and strive to become a demonstration or sample of the rural settlements in Suining and even across China. In January 2018, through field investigation, six scenic villages with distinctive features were selected as competition bases. The planning and design began in April and plans were appraised in August after completion. More than 600 teachers and students from 21 universities across China completed 106 sets of design projects within 4 months, and 56 sets of award-winning projects were finally selected. These planning designs will provide practical advice for specific construction work in the next step.

Contest Awards

Works Exhibition and Promotion

Participating Students and Teachers Visit Rural Bases to Conduct Field Research
(Source: Urban Planning Society of China)

Box 6-2 Professor Yang's Seven-Year Visit to Rural Areas to Guide Village Development

For seven consecutive years, Professor Yang Guiqing, Director of the Department of Urban Planning, College of Architecture and Urban Planning, Tongji University, has led teams of teachers and students to conduct rural field research and practice in the western mountainous area of Huangyan District, Taizhou City, Zhejiang Province. During the past seven years, Professor Yang and his teams visited Huangyan village more than 150 times and guided the construction in several locales such as the beach village of Yutou Township in Huangyan District and Wuyantou Village in Ningxi Town, which revitalized formerly declining villages. On the basis of these practices, Professor Yang led teams to condense a distinctive "new localism" planning theory, and the Rural Revitalization Methodologies was published in the Economy & Nation Weekly under Xinhua News Agency. Professor Yang's deeds were reported by many national media outlets, for example, as in the special report "Professor Going to the Countryside" on the CCTV News Probe. The seven-year rural investigation provided a lively "Huangyan sample" for the implementation of China's rural revitalization strategy.

Real Scene Photo of Villages in Huangyan: Dilapidated Houses Before Renovations

Real Scene Photo of Villages in Huangyan: Modern Functions Embedded After Renovations

Professor Yang Guides Rural Construction in Yutou Township, Huangyan District

Professor Yang Guides Rural Construction in Ningxi Town, Huangyan District

Bird's-eye view of Shatan Old Street, Shatan Village, Yutou Township, Huangyan District

(Source: CPC Committee of Yutou Township, Huangyan District, Taizhou City, Zhejiang Province, CPC Committee of Ningxi Town, Tongji University)

6.2 Poverty Alleviation

As the largest developing country in the world, China has always been an active advocate and powerful promoter of the world's poverty reduction cause. In the past 40 years of reform and opening up, China can claim tremendous achievements in the history of poverty alleviation and has contributed Chinese wisdom and Chinese programs to the global poverty reduction cause. By the end of 2018, the number of poverty-stricken people in rural areas in China[①] had decreased from 98.99 million at the end of 2012 to 16.6 million, and the incidence of poverty fell from 10.2% in 2012 to 1.7%. The number of poverty-stricken villages identified by the nation-wide data tracking had been reduced from 128,000 to 26,000, and 153 of the 832 poverty-stricken counties have declared being raised out of poverty. At present, China is mobilizing all social forces to participate in this crucial period of poverty alleviation, ensuring that the rural population living below the current poverty threshold and all impoverished counties are all lifted out of poverty by 2020 enabling problems of regional poverty to be solved.

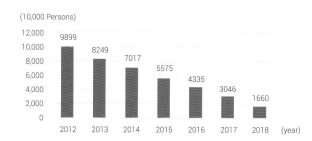

Figure 6-1 Rural Population Living Below the Current Poverty Threshold from 2012 to the end of 2018
(Source: according to statistical bulletin of the National Economic and Social Development over the years)

---

① By the rural poverty standard of 2,300 per year per person (constant 2010 RMB).

6.2.1 From Poverty Alleviation Through Regional Development to Targeted Poverty Alleviation

In the four decades of 1978-2018, China's rural poverty alleviation has shifted from a regional development in poverty-stricken areas to a targeted poverty alleviation approach for poverty-stricken households and population[②].

In the early stage of reform and opening up, a strong impetus for poverty alleviation through regional development was provided through the implementation of the reform of the land contract management system and the development of the rural commodity economy to release the policy's vitality, liberate the productivity, and stimulate the enthusiasm of the farmers. During this period, in response to the regional characteristics of poverty distribution, the Chinese government determined the focus of state support at the county level, which is the basis for implementing the poverty alleviation programs by region.

Beginning in the 1980s, the main strategy for poverty alleviation through regional development was to promote regional development in poverty-stricken areas and indirectly lift poverty-stricken people out of poverty. In 1982, the state launched the special poverty alleviation plan in "Three Western Regions" (i.e.Hexi and Dingxi in Gansu Province and Xihaigu in Ningxia Zhuang Autonomous Region), which kicked off a stage of organized, planned and large-scale poverty alleviation through regional development. In 1986, the

---

② Wang Sangui, Zeng Xiaoxi. From Regional Poverty Alleviation and Development to Accurate Poverty Alleviation: The Evolution of China's Poverty Alleviation Policy in the 40 Years of Reform and Opening up and the Difficulties and Countermeasures of Poverty Alleviation. Issues in Agricultural Economy, 2018 [08]:40-50.

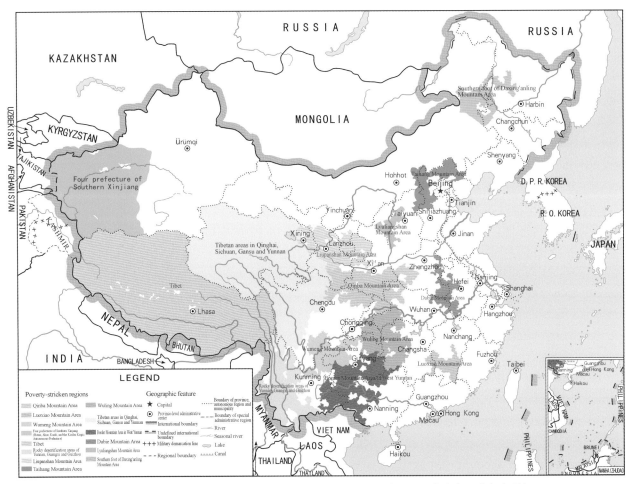

Figure 6-2　Distribution Map of Contiguous Poverty-stricken Areas and Areas Covered by Special Poverty Alleviation Policies in China
(Source: China Geological Survey of the Ministry of Natural Resources)

Figure 6-3　Cliff-top Villages in Liangshan, Sichuan Province turn into scenic spots (Photo by Rao Guojun)
(Source: http://www.cnsphoto.com)

state formulated poverty alleviation standards, set up institutions in charge of poverty alleviation, established special poverty alleviation funds, identified national poverty-stricken counties, and implemented development-oriented poverty alleviation actions that promoted regional poverty alleviation. After 2001, the state substantially increased the national poverty alleviation standards, adjusted poverty-stricken counties, and used contiguous destitute areas as the main battlefield for poverty alleviation through regional development in the new era, further consolidating the results of subsistence, and improving the overall quality of life of the poverty-stricken population[①].

Since 2012, the central government has taken poverty alleviation as its bottom line task and landmark indicator for building a well-off society in an all-round way and it has placed poverty alleviation in an important position in the governance of China during decision-making arrangements. It has also adjusted the previous model of poverty alleviation with the focus on regional development and taken targeted poverty alleviation as the basic strategy introducing a nationwide poverty reduction strategy of combining the programs of "Precise Poverty Reduction","Regional Poverty Reduction through Development" and "Social Guarantee"[②]. The main driving force for China's massive poverty reduction is economic growth, especially the continued growth of agriculture and rural economies in poor areas. Targeted development-oriented poverty alleviation has played an important role in helping the economic and social development of poverty-stricken areas by implementing regional infrastructure and public service construction, and helped to narrow the development gap between poverty-stricken areas and general areas. This targeted poverty alleviation further made the poor population the primary target of poverty alleviation through regional development, and greatly improved the effectiveness of poverty alleviation work.

6.2.2 National Poverty Alleviation Plan

Since 2012, to promote poverty alleviation more effectively, the state has introduced a series of guidelines and policies according to a basic strategy of targeted poverty alleviation, This will also ensure that the goal of poverty alleviation be achieved as scheduled. More than 100 policy documents or implementation plans have been issued by various ministries and commissions of the central government. All provinces, cities, counties and townships have formulated supporting documents according to their local conditions, consisting of such basic components as the institutional design, objectives and tasks, main strategies and safeguard measures.

In November 2015, the State Council issued the *Decision on Winning the Fight Against Poverty*, demanding to mobilize the power of the whole society and resolutely achieve poverty eradication. This policy proposes to "ensure that progress in development-oriented poverty alleviation is coordinated with overall economic and social development", "combine two targeted measures for poverty alleviation and developing contiguous poor areas with special difficulties, each through the application of strategies", "attach equal importance to poverty alleviation and ecological protection", and "integrating poverty alleviation with social security", etc.It is the general guiding principle and general deployment to win the fight against poverty, and it is the guiding document that runs through the whole process of poverty alleviation. In November 2016, the State Council issued the *National Plan for Poverty Alleviation in the 13th Five-Year Plan Period* to further strengthen atop-level design of poverty alleviation and achieve a seamless connection with national economic and social development planning. In order to further improve the top-level design and strengthen overall coordination, the Central Committee of the Communist Party of China and the State Council issued the *Three-Year Guidelines of Winning the Battle against the Poverty* in August 2018 as the action plan for the next three years. In 2019, the No. 1 Docu-

---

[①] The State Council Information Office, Press Conference on Issues on China's poverty reduction in 40 years.
[②] Report on China's Implementation of the Millennium Development Goals (2000-2015).

ment of the Central Government clearly identified poverty alleviation as the top priority, and proposed measures such as focusing on severely poverty-stricken areas and focusing on the development of long-term poverty alleviation industries.

6.2.3 Targeted Poverty Alleviation

The *Three-Year Guidelines of Winning the Battle against the Poverty* promulgated in 2018 emphasized "adhering to targeted poverty alleviation" as the basic strategy, and required that "targeted measures are implemented in terms of funding, projects, and recipients;every impoverished household is guaranteed help, every village has designated officials to carry out poverty eradication measures, and goals are met within the defined standards.Efforts shall be made to adapt to the local conditions and actual situations, identify the authentic recipients, makers, working approaches and exit measures in poverty alleviation, and to make substantial efforts in lifting the truly impoverished people out of poverty and truly reversing the impoverished state of China."

Through establishing the working mechanism of coordination by the central government, general responsibility assumed by the provincial governments and specific implementation by the municipal and county governments, targeted poverty alleviation conducts data tracking,accurately identifies the poverty-stricken population, selects the work teams, strengthens the frontline work force for poverty alleviation, increases investment in poverty alleviation, strengthens policy initiatives, provides a solid guarantee for poverty alleviation, and, thus, China's poverty alleviation work has entered a new stage of poverty alleviation. UN Secretary-General Antonio Guterres has said that the strategy of targeted poverty alleviation is the only way to help the poorest people achieve the ambitious goals of the *2030 Agenda for Sustainable Development*[①].

In recent years, in response to the call for targeted poverty alleviation, more and more social forces in China have actively participated targeted poverty alleviation work, becoming the backbone of poverty alleviation efforts and injecting a powerful force for targeted poverty alleviation. All parties involved accurately combine the actual situations of poverty alleviation counties in the fight against poverty and actively explore diversified paths for poverty alleviation.

6.3 Improving Rural Living Environment

Improving the rural living environment is an important task in implementing the rural revitalization strategy. In recent years, various regions and authorities in China have continued to intensify efforts to promote rural infrastructure construction and equal access to basic urban and rural public services. However, due to regional differences, the rural living environment is very uneven. In order to speed up rectification work, in February 2018, the Chinese government promulgated the *Three-Year Action Plan for Rural Human Settlement Environment Renovation*. The overall goal is to achieve a significant improvement in the rural living environment by 2020;the village environment is essentially clean and orderly, and the villagers' environmental and health awareness is generally enhanced. Among them, the primary task is to improve the basic environment, and then gradually improve the appearance of the villages, strengthen the villages' planning

---

① People's Daily. Targeted poverty alleviation, the greatest story written by China–Positive comments from international community on achievements of China's poverty alleviation efforts.

Box 6-3   Poverty Alleviation by Colleges and Universities——Tongji University's Efforts in Targeted Poverty Alleviation in Yulong County, Yunnan Province

Located in the westernmost part of the Dali Bai Autonomous Prefecture in Yunnan Province, China,Yunlong County is a national key county of poverty alleviation that features"ethnic groups, a mountainous region, poverty and a remote location." Tongji University has paired with Yunlong County since 2013 to engage in poverty alleviation, and has explored a special path with urban and rural planning as the core engine to promote targeted poverty alleviation in order to leap forward with social and economic development. The project was identified as one of the top ten typical projects of the Ministry of Education's third initiative of targeted poverty alleviation by universities directly under the Ministry of Education. By September 2018, 7 poverty-stricken villages of Yunlong County were removed from the list of poverty-stricken villages, and 24,787 impoverished residents had been lifted out of poverty.

Since 2013, Tongji University has made full use of the traditionally strong discipline of urban and rural planning, has organized teams to take the lead in carrying out the overall planning preparation via coordination and taking a leading role,has used top-level planning consultations to identified the roots for the poverty in Yunlong, and has pointed out the direction and focus for the urban and rural development and poverty alleviation in Yunlong County. Then, special planning was used to guide and promote the implementation of specific projects and establish a library of targeted poverty alleviation projects. Under the guidance of the planning, the Yunlong County Government concentrated on the development of the county seat of Caojian Town, focusing on strengthening the protection of ecological and cultural heritage, strengthening the construction of village and town infrastructure and public service facilities, effectively enhancing the service level of the county, promoting the development of surrounding villages, and improving the evironment of their habitat.

Since 2018, Tongji University has chosen Yong'an Village, a deeply poverty-stricken village, as a demonstration village to undertake the preliminary working philosophy of "planning and leading". The focus on poverty alleviation efforts has gradually deepened, going from the whole surface to single points. Efforts have been made to actively explore new ways of doing integrated urban and rural development and rural space optimization. Village planning was conducted to clarify the general direction of future development; planning was then used to coordinate the development process, and the project planning of comprehensive improvement in five aspects, such as road traffic improvement, residential environment improvement, rural community construction, rural industrial upgrading, and public space transformation. The process was carried out to form the series project library of Yong'an Village poverty alleviation demonstration site and promote Yong'an Village's poverty alleviation and rural revitalization.

In-depth Field Investigation in the Poverty-stricken Villages

Design Sketch of Villagers'Meeting Place
(Source: Tongji University, Shanghai Tongji Urban Planning and Design Institute Co., Ltd.)

Box 6-4 Poverty Alleviation Through Science and Technology—Poverty Alleviation Efforts of China Association for Science and Technology in Lanxian County, Shanxi Province

Lanxian County, Lvliang City, Shanxi Province is one of the 36 key contiguous poor counties identified by the state and one of the national-level poverty-stricken counties for poverty alleviation through regional development as well as the selected site for poverty alleviation by China Association for Science and Technology. Changmen Village is located in Wangshi Township, Lanxian County, and is a poverty-stricken village supported by the China Association for Science and Technology. Beginning in 2017, according to the unified deployment of the China Association for Science and Technology, the Urban Planning Society of China actively carried out poverty alleviation through science and technology, and organized relevant professional and technical teams to jointly assist Changmen Village in the targeted poverty alleviation and poverty alleviation actions. Through the creation of a platform, the Urban Planning Society of China has established a bridge for poor rural communities to communicate and cooperate with China's top technical forces. In the past two years of work, the aid teams from Shanghai and Shanxi have compiled a plan for the implementation of the project on the basis of in-depth investigation and research, solicited opinions from the villagers, and publicized it in Changmen Village in May 2019. The plan was adopted by a show of hands at the villagers' representative meeting.

On the basis of considering the topography, ecological sensitivity and planting characteristics of the village, the spatial resources and industrial planting are integrated considering regional market demand. Therefore, in accordance with

Planning of Festival Activities | Folk Culture | Village House Holiday | Educational Activities | Picking Experience and Forest Activity

Month	Current Activities of Farmers	Planned Festival Activities
January	farm slack season	viewing winter scenery, photography activities
February	cleaning latrine pit, composting, celebrating lunar festivals	rural new-year celebration activities, potato banquet
		Chinese New Year Lantern Festival
		dough figurines, window paper cutting
March	plowing	agricultural education
April	crop germinating, thinning out, hoeing	learning folklore (paper cutting)
May	hoeing	historical and cultural visits (village culture, red culture)
June		summer mountain walk
July	farm rest season	viewing potato flowers
		summer leisure activities
August	autumn harvest: millet, coarse rice, sorghum, potatoes	harvest Festival
		summer school for learning agriculture, holiday sketch
September	autumn harvest: maize	Mid-Autumn Festival activities
		field picking and sale of agricultural products in autumn harvest
October	crop threshing on the threshing ground	village leisure activities for National Day holiday
November	farm rest season	operation of rural home inns and tourism Minshuku
December		training for villagers

Planning of Village Festival Activities

the distribution of topography and water sources in the village area, crop planting zoning guidelines are proposed with ecological priority; and the integration and upgrading of the industrial chain is carried out according to the resource characteristics of each region; meanwhile, festival activities are planned to promote the integration and upgrading of village industries according to seasonal characteristics of village agricultural planting.

In order to further deepen and implement the planning visions, the China Association for Science and Technology, the Urban Planning Society of China and its Academic Committee for Rural Planning and Construction organized more than 20 multi-disciplinary experts and scholars to conduct on-the-spot investigations and extensive in-depth discussions on the industrial structure, ecological restoration, and living environment construction in Changmen Village. In this process, they regarded the poverty-stricken efforts in Changmen Village as long-term work and it has shifted from "poverty alleviation" to "enhancement of intelligence". Under the continuous promotion of "Poverty Alleviation through Science and Technology" by the China Association for Science and Technology, the county has been lifted out of poverty.

The First Batch of Experts Conducted Field Research in 2017   The Expert Team Entered the Village for Discussion in 2019

(Source: Urban Planning Society of China)

management and improve construction and management mechanisms. Under the guidance of this policy, renovations of rural living environments have been comprehensively promoted nationwide.

6.3.1 Improving Rural Housing Conditions

Housing security is the basic goal of improving the rural living environment. The renovation of China's dilapidated houses in agricultural areas originated from the damage and collapse of a large number of rural houses caused by the extraordinary snowstorm disaster in the mountainous areas of southern China in 2008. After the disaster, Guizhou Province took the lead in launching the renovation project of rural dilapidated buildings in China, and successively moved villagers living in old and dilapidated houses into new homes through government subsidy funds and grants.

On the basis of supporting the pilot project for the renovation of dilapidated buildings in Guizhou Province, the central government extended the subsidy to serve dilapidated houses in nearly 800,000 rural households in 950 counties in the central and western regions. With the continuous expansion of the pilot's scope, full coverage of rural areas was achieved in 2012. Since 2013, various authorities across China have continued to intensify the renovation of dilapidated houses in rural areas, improve policy measures, and strengthen guidance and supervision. In 2018, in conjunction with the task of "housing safety and security" in the overall goal of poverty alleviation across China, the Ministry of Housing and Urban-Rural Development, along with the Ministry of Finance and the State Council Leading Group Office of Poverty Alleviation and Development,strengthened the deployment for renovation of rural dilapidated buildings for key targets such as poor households. According to statistics from the Ministry of Housing and Urban-Rural Development, as of March 2019, more than 6 million poor households recorded in the data tracking have completed the renovation of their dilapidated buildings.

On the basis of solving the problem of rural housing safety, in recent years, various provinces, cities and regions have proposed many local promotions and management policies for the improvement of rural housing construction and rural styles.For example, the *Implementation of Opinions on Concretely Improving the Construction and Management of Rural Housing* (2017) of Zhejiang Province and the *Notice on Further Promoting the Reporting and Management of Rural Housing Construction Planning*(2018) of Hainan Province. Beginning in 2019, the pilot projects of rural housing construction started nationwide, aiming at improving the design and service management level of rural housing construction, and building a number of livable model rural houses with modern functions, local styles, cost-effectiveness, structural safety and green environmental protection. The main content of this work is to explore the mechanism to support farmers in building livable farmhouses, to explore the mechanism of organizing farmhouse design forces to go to the countryside, and to establish a rural construction craftsmen training and management system to comprehensively promote the further upgrade of rural housing's quality and living environment.

6.3.2 Improving Rural Infrastructure Levels

Since the end of the 20th century, China has started a large-scale systematic rural infrastructure transformation and construction project, including roads, electricity, drinking water, telephone networks, radio and television, and the Internet, which is called the Extending Radio and TV Broadcasting Coverage to Every Village Project. The corresponding administrative authorities have issued a series of policies to accelerate the construction. Over the years, the transformation of rural infrastructure has provided the basic conditions for the development of rural areas across China.

Since 2014, China has carried out the construction of "Four Good Rural Highways", which means to"build good highways, maintain good management, keep good protection and maintain good operation", and the goal is to complete all paved roads in townships and villages by 2020, and all the funds for maintenance will be included in the fiscal budget. All the eligible administrative villages meeting the conditions will be provided bus access and, in essence, the rural logistics network covering the county, township and village

Box 6-5   Rural "Three Reforms" Plan in Danzhai County, Guizhou Province

Beginning in 2017, Danzhai County, Qiandongnan Miao and Dong Autonomous Prefecture, Guizhou Province, implemented the "three reforms" plan for the renovation of rural dilapidated houses and attached kitchens, toilets and livestock pens and poverty-stricken households, low-income households and deeply impoverished areas, the key targets for the "three reforms". The implementation of the project has effectively helped the people with the most hardship in rural areas solve the basic housing safety problems such as bedroom and kitchen separation, toilet and livestock pen separation, and separation of people and animals, thus ensuring the health of the rural people and improving the living environment of the villages.

Brand New Miao Village After Renovation

A Villager Rests in Front of the Renovated New House

Workers Are Carrying Out Renovation of Livestock Pens

Workers Are Carrying Out Renovation Work in Dilapidated Buildings

Construction of Rural dilapidated Housing and "Three Reforms" in the Yaohuo Village, Xingren Town, Danzhai County, Qiandongnan Miao and Dong Autonomous Prefecture, Guizhou Province (Photo by Huang Xiaohai)
(Source: http://www.cnsphoto.com)

Box 6-6   Uniform Planning and Construction of Centralized Resettlement Village in Xianju County, Zhejiang Province

Shuiduitou Village, Butou Town, Xianju County, Zhejiang Province is a concentrated resettlement village for migrants from mountainous areas. The new village adopts unified planning, unified design and unified construction. The farmhouse-style conjoined houses of the new village are regularly arranged, the roads in the village are flat and spacious, and the trees are lined in front of and behind the houses. The distant mountains and local waters add luster to this garden-style village, which has won honors such as the beautiful livable demonstration village in Zhejiang Province, the demonstration village for rural housing reconstruction, the characteristic farmhouse village, and the forest village. In recent years, the village has begun to focus on the development of farmhouse and leisure tourism projects, and used the beautiful and livable environment to promote rural revitalization.

Garden-style New Village in Shuiduitou Village, Butou Town, Xianju County, Zhejiang Province (Photo by Hua Wenwu)
(Source: http://www.cnsphoto.com)

levels will be built. From 2018 onwards, this work has entered a new stage of high-quality development at the height of implementation of the rural revitalization strategy and winning the battle against poverty. It is focused on outstanding issues, and further transferred the focus from road construction to improving the road management and operation policy mechanism. After five years of promotion, the newly built rural roads amounted to 1.392 million kilometers, and in total rural roads reached 4.05 million kilometers in length with the proportion of villages and towns with paved roads reaching 99.5%. The transportation network has been initially formed with the county seats at its center, the townships as intersections, and the administrative villages as the branches. Even the relationships between the villages and urban and rural areas have become closer.

In order to narrow the gap between urban and rural areas and promote rural modernization, the governance of rural human settlements has gradually begun to pay attention to the treatment of rural domestic garbage, sewage and toilet manure, specifically the "toilet revolution" for toilet manure treatment. This action originated from the construction and management of

Box 6-7   Big Data Information Platform of "Transport to Every Village Project" in Guizhou Province

In order to solve the problem of "the difficulty of traveling for the people, the difficulty of returning to school for students, and the difficulty of rural freight" in Guizhou's mountainous rural areas, the rural road traffic platform under the "Transport to Every Village Project" was launched in July 2017, and the vehicle network and precise pairings were realized with the help of big data. After more than a year of hard work, through the integration of rural passengers; freight, travel and logistics needs;safe, convenient and efficient travel services;logistics services including shuttle buses, chartered cars, online car-hailing, taxis, buses, student-customized shuttles, ticket purchases and express delivery into the village; small-article courier services; e-commerce logistics; and rural freight transportation, are all provided to villagers. The plan has reduced the cost of rural passenger transport operations, facilitated the travel of the rural population, opened up the last mile of rural logistics, provided a solid foundation and guarantee to e-commerce for outgoing transport of local commodities and goods from the mountain areas, and, finally, has provided a new model for poverty alleviation in terms of traffic.

tourist toilets initiated by the National Tourism Administration across China in 2015, whose overall results have been fully affirmed and was promoted in 2017 as a specific work of the rural revitalization strategy. In 2018, the No. 1 Document of the Central Government clearly adhered to the rural "toilet revolution" and vigorously carried out the construction and renovation of rural sanitary latrines.They simultaneously implemented the treatment of manure, encourage d localities to combine toilet waste and livestock and poultry farming waste and recycle them. Since 2018, nearly 24,000 new tourist toilets have been built, renovated and expanded, and about 15,000 tourist toilets have been built in the central and western regions. At present, this work has become the focus of all levels of government in promoting the construction of people's livelihood projects. In 2019, the central government planned to allocate 7 billion yuan, aiming to promote the renovation of toilets for about 10 million farmer households in 30,000 villages.

6.3.3   Protecting Rural Historical and Cultural Heritage

As early as the beginning of reform and opening up, Guizhou Province of China has already begun to engage in the protection of ethnic villages, becoming the first province in China to begin the protection and utilization of rural historical and cultural heritage. In the 1990s, Guizhou Province began to introduce the theory of international eco-museums, and took the lead in establishing eco-museums in China, such as Longjia Village in Liuzhi, Zhenshan Village in Huaxi, Longlisuo in Jinping, Tang'an Village and Dimen Village in Liping, and etc. The construction of the eco-museums has further strengthened the status of the villagers as cultural masters; cultural equality, autonomy and maintenance have become distinct themes. In 2005, Guizhou Province introduced an international cultural landscape theory to guide the protection of cultural relics. In October 2008, the International Symposium on the

Box 6-8   Renovation of Rural School Toilets in Zhijin County, Bijie City, Guizhou Province

Zhijin County, Bijie City, Guizhou Province is the only county in Guizhou undertaking the project jointly organized by the Ministry of Education of China and UNICEF. With the support of the project, in 2018, Zhijin County carried out toilet renovations for 10 rural schools. The new toilet replaced the previous messy, dirty and disorderly conditions and created a good toilet environment for teachers and students, and also accumulated experience for Zhijin County to engage in the in-depth development of the "Toilet Revolution".

Pupils of Qimo Primary School, Qimo Sub-district Leave After Using the Toilets (Photo by Qu Honglun)

Clean and Tidy Toilet Yina Primary School, Yina Town After the Renovation (Photo by Qu Honglun)

Women's Toilet in Yina Primary School, Yina Town (Photo by Qu Honglun)

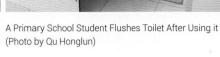

A Primary School Student Flushes Toilet After Using it (Photo by Qu Honglun)

(Source: http://www.cnsphoto.com)

Protection and Sustainable Utilization of Village Cultural Landscapes was held and produced the "Proposal for the Protection and Development of Village Cultural Landscapes", which has received great attention in the field of cultural heritage at home and abroad.

Based on local practice and international visions, the protection of rural historical and cultural heritage in China is one of the important vehicles for "traditional

Box 6-9 Traditional Chinese Village Digital Museum

In 2017, the General Office of the Ministry of Housing and Urban-Rural Development issued the *Notice on Doing a Good Job in the Construction of Excellent Village-based Museums under the Traditional Chinese Village Digital Museum* (Letter No.137 [2017] of the General Office of the Ministry of Housing and Urban-Rural Development on Rural Affairs), and officially launching the construction of digital museums in traditional Chinese villages. After one year of hard work, the first phase of the digital museum development and construction tasks and the construction of 165 villages were completed and officially opened online on April 28, 2018. The Digital Museum is divided into a general exhibition hall, individual village-level halls and a panoramic roaming mobile phone client. The main exhibition hall on the site has such column options as list, exploration, academics, activities, cultural creation, community, information, and others. It is a digital platform for the encyclopedic and panoramic display of traditional villages and a platform for the exchange of the academic resources of traditional villages.

Official Website   Mobile Client

(Source: Official Website of Traditional Chinese Village Digital Museum)

Chinese villages". After years of exploration, a set of policy systems has been formed and the corresponding protection has become a substantially important component of policy issues related to rural developmentin China. "Traditional Chinese villages" refers to "villages with material and non-material forms of cultural heritages and high historical, cultural, scientific, artistic, social and economic values". Since 2012, relevant investigations, assessments and documentation have been carried out nationwide, and corresponding planning and management policies have been formed. In 2014, the central finance committee has started to support protection projects of traditional villages, and has increased investment over the years. Since 2016, China has stepped up policy guidance on the maintenance and promotion of the effectiveness of traditional village protection, launched a warning and exit mechanism, promoted the construction of the "Traditional Chinese Village Digital Museum", and held the International Convention on the Protection and Development of Traditional Villages. By 2018, the list of "traditional Chinese villages" is divided into five batches, totaling 6,799, accounting for about 1.2% of the total number of administrative villages in China. Among these traditional villages, the proportion of villages supported by the central finance committee is about 54%.

6.4 Small Cities and Towns Construction and Characteristic Development

6.4.1 Development History and Roles of Small Cities and Towns

The roles of small cities and towns in China's urbanization process cannot be underestimated. The number of small cities and towns is extremely large in China and the importance of small cities and towns has been emphasized in the national and urban development strategies of different periods. From the *Small Cities and Towns, Big Problems and the Grand Strategy* in the 1980s to the *Promotion of the Coordinated Development of Large and Medium-sized Cities with Small Cities and Towns* after 2000, small cities and towns have always played an important role in China's urbanization and integrated urban and rural development.

From the perspective of historical evolution, the development of small cities and towns since the 1980s can be divided into three stages: the first stage was the recovery period (1979-1984), when the rural economy witnessed a gradual recovery from the relative stagnation of the planned economy period through the reform of rural household responsibility system, the transition from the people's commune to the township system through the zoning adjustment and reform, and

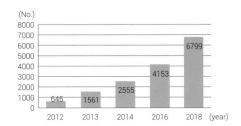

Figure 6-4 Total Number of Traditional Chinese Villages by Batch
(Source: Ministry of Housing and Urban-Rural Development, China)

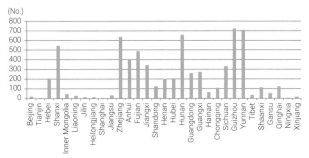

Figure 6-5 Number of Traditional Chinese Villages by Provinces and Municipalities (not including HongKong, Macao or Taiwan)
(Source: Ministry of Housing and Urban-Rural Development, China)

Box 6-10　Yuelai Town, a Small Town in Central Jiangsu, China

Yuelai Town is located in the southeast of Haimen City, connecting with Linjiang New District in the south, Haimengang New District in the north, Changle in the west, and 30 kilometers away from Haimen City, and the total area of the town is 141.44 square kilometers. The area under the jurisdiction of Yuelai Town is the birthplace of China's national modern industry, especially light industry such as textiles and clothing, which dates back to a long time ago and is still the leading industry supporting the development of small cities and towns in the region. Yuelai Town has been known as the hometown of leather shoes. However, in recent years, with labor-intensive industries shifting to the hinterland, local governments have also experienced problems such as the difficulty of recruiting workers. Yuelai Town is actively introducing new industries, upgrading its existing traditional manufacturing industries, and promoting transformational development, such as the introduction of the Eastern Educational Equipment Innovation Industrial City, combining manufacturing and educational equipment experiences; and introducing emerging manufacturing industries to the area such as precision instrument processing. The development history of Yuelai Town is the epitome of the development of small cities and towns in the south of the Yangtze River.

Real Scene Photo of Industrial Cluster in Yuelai Town

Real Scene Photo of Yuelai Town Seat

Eastern Educational Equipment Innovation Industrial City Renovated from Former Yuelai Middle School
(Source: College of Architecture and Urban Planning, Tongji University)

the change of small cities and towns from point-based management units to planar regional administrative units with management targets covering a large number of villages; the second stage was reform and take-off period (1984-1997), when the reform of the household registration system in 1984, the reform of the social enterprise system, the reform of administrative divisions, and the initial changes in the economic system, accelerated the transfer of the rural population to small cities and towns, as well as the rapid development of township enterprises. The third stage was the adjustment period (1997-present) when,with the rapid implementation of the national land management system, the improvement of the fiscal and taxation systems, and the reform of the housing market, resulted in the acceleration of urban development and the decline of the status of small cities and towns. With the introduction of the new-type urbanization strategy, small cities and towns have been repeatedly mentioned in relevant government documents as nodes linking cities and villages, and the development status of small cities and towns has thus been further valued.

In recent years, the role of small cities and towns has been further expanded, and people from all walks of life have gradually realized the role of small cities and towns in serving as rural fulcrums. At the same time, developed provinces have a further understanding of the roles of small cities and towns in economic growth and transformation. As of the end of 2016, there were 21,116 administrative towns and 10,529 townships in China.

6.4.2 Characteristic Development of Small Cities and Towns

In 2014, Zhejiang Province first proposed the issue of development and construction of characteristic towns. In April 2015, Zhejiang Province issued the *Guiding Opinions on Accelerating the Planning and Construction of Characteristic Towns*, which defines that "the characteristic town is a development space platform that is relatively independent of the urban area and has clear industrial positioning, cultural connotations, tourism and certain community functions." In June of that year, the list of the first batch of characteristic towns in Zhejiang Province was announced. At the end of 2015, the main state leaders indicated in a report to the characteristic towns that "the construction of characteristic towns and small cities and towns is promising."

In February 2016, the State Council issued the *Several Opinions on Deepening the Construction of New Urbanization*, emphasizing the acceleration of the cultivation of small and medium-sized cities and characteristic small cities and towns; and in July 2016, the Ministry of Housing and Urban-Rural Development, the National Development and Reform Commission and the Ministry of Finance jointly issued the *Notice on Initiating Development of Characteristic Small Cities and Towns*, which determined that by 2020, about 1,000 characteristic towns will be cultivated, which will drive the building of small cities and towns across China. In October 2016 and January 2017, the Ministry of Housing and Urban-Rural Development and the China Agricultural Development Bank and the National Development Bank issued policies to promote policy finance and development finance to support the construction of small cities and towns, and further expanded the focus of financial support to the construction of a livable environment in small cities and towns and the construction of traditional cultural heritage. In February 2017, the National Development and Reform Commission joined the National Development Bank to issue a

Box 6-11   Guyan Painting Township—Dagangtou Town, Liandu District, Lishui City, Zhejiang Province

Guyan Painting Township is one of the first eight characteristic towns in Zhejiang Province that were selected as part of the first batch of characteristic towns, a national AAAA scenic spot, a national water conservancy scenic spot, the world's first irrigation project heritage and a UNESCO heritage site. The construction of characteristic towns is guided by the development of scenic spots, follows the principle of combining government leadership and market operation, and strives to form a development path that is mainly driven by the market and supplemented by the government. Such a development model has played a positive guiding role in promoting local development. At the same time, it also faces inevitable problems: cooperation and competition between industrialization, large-scale operation and artistic production components; the hugely different levels of comprehensive development and management of cultural tourism resources with that of developed countries; and, lastly, unfavorable tourist experiences. At the same time, the scenic area management committee was introduced to seek common development with the local towns and villages, but its existence also generates games and contradictions between the management committee and the local development in terms of appeal and benefit distribution. These problems must be faced in the development process of Guyan Painting Township, and they are also the sources of power that urge a continuous upgrade and optimization of the development path of small towns.

Natural Landscape of Guyan Painting Township

Township Scene of Guyan Painting Township

Barbizon Oil Painting Exhibition and Entertainment Event Planning in Guyan Painting Township
(Source: College of Architecture and Urban Planning, Tongji University)

policy, proposing to adopt the development of financial support for construction of characteristic towns as one of the main tools for poverty alleviation.

The above series of policies and action plans, as well as the subsequent characteristic town construction action plans formulated by various provinces and cities, have ignited a wave of "characteristic town fever", and local governments across China have successively launched the fostering and creation of special towns. In October 2016 and August 2017, the Ministry of Housing and Urban-Rural Development and other central ministries or commissions jointly selected and published a total of 403 characteristic towns to be fostered or created. So far, there have been two models for the construction of characteristic towns. The first one is the industrial town model represented by Zhejiang with the focus on promoting the upgrade of traditional industries and nurturing a new economy. The other is the characteristic development model of small cities and towns, which means to promote the development of local characteristic industries, absorb the rural labor force and promote local urbanization, drive the improvement of village and town infrastructure, improve the overall livability level, and protect local traditional culture, thus promoting the comprehensive and coordinated economic, social and cultural development of small cities and towns.

In December 2017, the National Development and Reform Commission, the Ministry of Land and Resources, the Ministry of Environmental Protection and the Ministry of Housing and Urban-Rural Development jointly issued the *Several Opinions on Standardizing the Construction of Characteristic Towns and Characteristic Small Cities and Towns*, which clarified the conceptual differences between characteristic towns and characteristic small cities and towns and put forward five basic principles for standardizing the promotion of characteristic towns and small cities and towns in various regions: adhere to innovation and exploration, adhere to adaption to local conditions, adhere to industrial construction, adhere to a people-oriented development approach and adhere to market domination. On August 30, 2018, the National Development and Reform Commission issued a document entitled the *Notice of the General Office of the National Development and Reform Commission on Establishing a High-Quality Development Mechanism for Characteristic Towns and Characteristic Small Cities and Towns*,

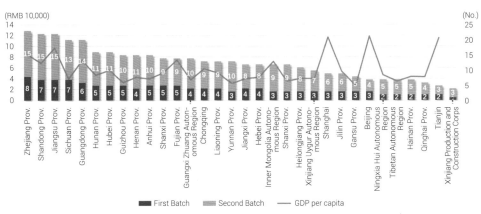

Figure 6-6 Comparison of the Number of Characteristic Towns and Economic Development Levels in Different Provinces and Cities
(Source: Zhang Li, Bai Yuxin. Characteristics Analysis and Discussion of 403 Small Cities and Towns in China [J]. Development of Small Cities &Towns, 2018, 36 (09): 20-30.)

Comparison of the Scale of Towns Across China and 403 Characteristic Towns   Table 6-2

Index	Towns Across China		403 Towns		Proportion of 403 Towns Above the National Average
	Average	Median	Average	Median	
Permanent Population of Town Jurisdictions (10,000 persons)	4.15	3.46	5.66	4.46	54.4%
Permanent Population of Town Seats (10,000 persons)	1.08	0.65	2.66	1.77	66.7%
Area of Built-up Area (hectare)	219.5	139.4	441.5	162.0	44.7%
Per Capita Construction Land Area (m^2/person)	240.7	198	234.7	124.2	25.1%

(Source: Zhang Li, Bai Yuxin. Characteristics Analysis and Discussion of 403 Small Cities and Towns in China [J]. Development of Small Cities &Towns, 2018, 36 (09): 20-30.)

which gives standardization and rectification to the unreasonable phenomena in the process of creating characteristic towns, and encourages the small town development model with distinctive features, integration of industries and towns, and a focus on an ecological environment.

6.5 China's Rural Practice

6.5.1 Beautiful Villages: Local Practice for Improvement of Human Settlement Environment

(1) Construction of Beautiful Villages in Zhejiang

The construction of beautiful villages in Zhejiang originated in 2003. Xi Jinping, then secretary of the Zhejiang Provincial Party Committee, creatively proposed and implemented the "Demonstration by a Thousand Villages and Renovation of Ten Thousand Villages" project, which opened a new historical journey of Beautiful Villages and Beautiful China. During the construction of Beautiful Villages in the past 20 years, Zhejiang has accumulated rich experience. Among them, the introduction of many innovative policies, such as land policies, fiscal policies, and village creation policies in the scenic spots, has effectively promoted the resolution of many problems in rural areas. All cities and counties adhere to the beautiful rural construction path according to local conditions, such as the construction of the Cultural Hall in Lin'an, the protection policy of Songyang Ancient Village, the Suichang Farmhouse Complex, the Jiangshan Village Song Culture, the Anji Standardization Construction, and the Deqing System Reform and Innovation. In the process of rural industry development, the active support or full participation of relevant enterprises not only provides a case base for enterprises to participate in rural construction, but also facilitates the local governments to accurately attract investment. For example, Yuanjian Group made plans for a beautiful village and "New Garden," Zhejiang Construction Investment Environment Engineering Co., Ltd,vowed to"become the nanny of rural sewage treatment", and Zhejiang Tourism Group "used the ancient village protection and utilization fund to reshape nostalgia", to name a few.

(2) Jiangsu's Construction of Characteristic Rural Villages

In order to thoroughly implement the decision-making and deployment of the central government on urban and rural construction and the work of "agriculture, rural areas and farmers", we must seriously infer the principles within a series of important speeches and new concepts, new ideas and new strategies from within governance of China under General Secretary Xi Jinping. Furthermore, we must strive to

Box 6-12 Zhejiang Anji Mode

Anji County is a typical mountainous county. After experiencing the pain of industrial pollution, in 1998 Anji County abandoned the road of an industrialized county, and in 2001 proposed the development strategy of becoming an ecological county. In 2003, Anji County, in conjunction with the "Demonstration by a Thousand Villages and Renovation of Ten Thousand Villages" project of the Zhejiang Provincial Committee, implemented the "joint-double project" with the content of "Demonstration by ten paired villages and the Renovation of a hundred paired villages" in the county and adopted various forms to promote rural environmental remediation. On this basis, Anji County took the lead in the construction of "China's Beautiful Villages" in 2008 and used it as an important vehicle for the new round of development. It is planned to build the administrative villages of the whole county into beautiful villages that feature not only the beauty of the village, but entrepreneurship, harmony, and happiness for all through "industry improvement, environmental improvement, quality improvement and service improvement" within 10 years. Based on the advantages of its local ecological environment and resources, Anji is vigorously developing emerging industries such as the bamboo tea industry, ecological rural leisure tourism, bio-pharmaceuticals, green food, new energy and new materials with the overarching concept of good village management.

Xiayang Village, Anji County, Zhejiang Province
(Source: Shanghai Tongji Urban Planning & Design Institute)

Box 6-13 Jiangsu's Characteristic Rural Village Construction

Xinghua of Taizhou City has established its own positioning as a "livable garden waterside town" and outlined its characteristics with five key elements: "water, green, shoaly land, earth and literature". The 6provincial-levelpilot villages in Xinghua have adapted to the conditions of the villages, gave full play to the role of local craftsmen, highlighted the original features of the villages, and showcased the elements of Xinghua through the colored drawing of farmer paintings, wooden fences and bridges, and water reeds. At the same time, they proposed not to cut down trees or backfill the river, and to demolish fewer houses, prohibit the whitewashing of houses, highlight the localization of building materials, make good use of old bricks and tiles, and use old objects to make materials on the spot, making do with what they already have, and trying to retain the original local characteristics to the utmost extent. In terms of industrial development, Xinghua created a characteristic agricultural business card represented by mass planting of cauliflowers, Xinghua Rice and Xinghua Hairy Crab. Xinghua regarded creative agriculture as a new engine for the development of leisure agriculture and a composite of the integrated development of primary, second and tertiary industries, focused on the development of culture-type, eco-creative agriculture, mined agricultural cultural heritage, enriched cultural and creative products connected to Zheng Banqiao and Water Margin, and cultivated creative agricultural products and creative farmland landscapes. Dongluo Village relies on the scenic area for the massive planting of cauliflowers to create the Flower Sea of Four Seasons, and is committed to "planting good scenery" and "selling good scenery". At the same time, it has cooperated with Vanke to build the "88 Warehouses" brand of agricultural and sideline products, and also planned the "online + offline"marketing of Xinghua hairy crab and osmanthus wine.

Longwang Temple, Meilin Village, Xixiashu Town, Changzhou City, Jiangsu Province
(Source: Shanghai Tongji Urban Planning & Design Institute)

improve the construction level of socialist new rural areas and constantly consolidate the foundation for building the New Jiangsu with a strong economy, rich people, a beautiful environment and a high degree of social civilization with the new practice of "Two Focuses and One High Level" (the focus on the quality of talent training and the focus on the construction of a contingent of teachers alongside the establishment of a high level of an applied talent training system).In the end of June 2017, the Jiangsu Provincial CPC Committee and the Provincial Government officially issued the *Action Plan for the Construction of Characteristic Rural Villages in Jiangsu Province* and the *Pilot Program for the Construction of Characteristic Rural Villages in Jiangsu Province*,starting the construction of characteristic rural villages at the provincial level to create "characteristic industries, characteristic ecology, characteristic culture, shaping rural scenery, rural architecture, rural life, building beautiful villages, livable villages, and vibrant countryside" and this showcases the true views of the beautiful rural villages of Jiangsu that feature"excellent ecological environments, beautiful villages, special industries, rich farmers, strong collective economies, and good rural customs". All counties and cities have actively engaged in the construction and have thus accumulated a lot of experience.

6.5.2 Industrial Revitalization: New Practice of Comprehensive Development of Rural Industry

(1) Internet - Taobao Villages

The booming development of Taobao Villages in China has benefited from the widespread popularity of Internet e-commerce in rural areas. A network cable has connected the national and global markets, and has also supported the dream of farmers for wealth. With the further development of Internet technology and continuous iteration,new technologies such as mobile smart devices and new live broadcast platforms are also profoundly reshaping the development model of the e-commerce industry of Taobao Villages.

Xiangzhai Village, Houji Town, Zhenping County, Nanyang City, Henan Province, known as the "First Village of China Koi", hit the Internet with the "koi wave"in 2018, and successfully ranked among "China Taobao Villages". However, such a village known for its koi culture is not a water-rich land of fish and rice, but a traditional farming area located on the northern edge of the Nanyang Basin. The development of the local koi culture industry has been an experience or process of starting from scratch. In the late 1970s, the first goldfish farmer appeared in the village. From this 30-square-foot fish pond, the goldfish culture industry in Houji Town started. In 1983, the villager named Li Guangzhi introduced a batch of goldfish to promote sales in the market, and the profits were very impressive. The first batch of goldfish farmers abandoned land farming and dug ponds to engage in the professional production of goldfish farming.

After nearly two decades of development, Houji Town, to which the village of Xiangzhai Village is subject, was officially named the "Hometown of Chinese Goldfish" by the Organizing Committee of the Local Township Products of China. Around 2006, the villagers began to gradually develop koi with higher profit returns. The koi culture in the village is mainly based on the self-cultivation of the villagers and the large-scale farming of the companies. Changyan Aquarium Culture Co., Ltd. was established in 1988 and it is the largest local Koi breeding enterprise in Houji Town. Based on Changyan Aquarium Culture Co. Ltd.'s high-quality resources,the Zhenping County Government initiated the establishment of the

Runfa Ornamental Fish Farmers Cooperatives Association so the villagers could cooperate with the company to carry out goldfish farming, sales, or work in the company. At the same time, the organization is also an e-commerce incubator base and Changyan Koi Poverty Demonstration Base, which is organized and operated by the Luoyang Shanxun Group, a service provider cultivated by Taobao Village and commissioned by Zhenping County. The Shanxun team has expanded the sales market of Koi by training local villagers to master e-commerce skills, which has reduced the sales pressure of farmers and increased the income of villagers.

In 2018, Xiangzhai Village was elected "China Taobao Village" of the year. By 2018, the production of Koi in Houji Town accounted for 60% of the national total, of which the production of Koi in the village was 40% of China, and there were more than 300 farmer households, with an annual output value of more than 80 million yuan, with sales reaching coastal developed areas in China, as well as Japan and Hong Kong and some countries and regions in Southeast Asia.

(2) Rural Tourism - Yuanjia Village

Yuanjia Village, Liquan County, Xianyang City, Shaanxi Province, China, is located in the hinterland of the Guanzhong Plain. In recent years, through the development of local tourism, the village has won the honorary title of being one of "China's Top Ten Beautiful Villages", "Traditional Chinese villages" and "Chinese Charm Villages". It has been called the "Lijiang of Shaanxi" and it successfully created what is known as the "Yuanjia Village" mode of development.

Yuanjia Village has experienced the following stages in its development from a small village to a famous rural tourist destination: hard work in the 1950s and 1960s, which turned barren places into a more prosperous development zone; reform and opening up in the 1970s and 1980s when township and village enterprises flourished and set up "five small industries" (Note: this generally refers to five small industrial enterprises such as small steel enterprises, small coal mine enterprises, small machinery enterprises, small cement erprises, and small chemical fertilizer enterprises); after 2000, when the high pollution enterprises were shut down due to strict pollution control, leading to the gradual decline of Yuanjia Village; villagers had to leave home to find jobs, and there was a "hollowing" phenomenon. In 2007, rural tourism began to develop, forming a workshop street and a snack street. After 2011, the village combined folklore, creative culture and art blocks with the promotion of Guanzhong folk culture at its core, and

Figure 6-7　Fish Pond in the Village
(Source: Nanjing University Space Planning Research Center)

Figure 6-8 Changyan Ornamental Fish Breeding Base
(Source: Nanjing University Space Planning Research Center)

Figure 6-9 Xiangzhai Village E-commerce Incubation Base
(Source: Nanjing University Space Planning Research Center)

Figure 6-10 Villagers' Private Breeding and Selling of Ornamental Fish
(Source: Nanjing University Space Planning Research Center)

relying on the functional supplements stemming from urban requirements, they introduced external commercial projects for market operations, thus successfully shaping the "Yuanjia Village" model.

The success of Yuanjia Village lies in the persistence of both the collective economy and the achievement of common prosperity. China's clan concept and local culture have determined that the village is based on a nostalgic governance model and the development direction of neighbors ought not to be much different than another's. This is the foundation on which China can achieve a flourishing spirit and the concept of Yuanjia Village's common prosperity. Under the leadership of the village party branch and the village committee, all villagers are mobilized to participate in the development of rural tourism. For example, a traditional food can be developed by individuals who have the ability to develop it. After the brand is formed, a company will be formed and shares will be transferred among the villagers. The villagers and merchants willing to join in will voluntarily share the proceeds, which means "every family has business, and everyone can be employed." By adjusting income distribution and redistribution, polarization can be avoided and common prosperity can be achieved. This approach not only greatly improved the enthusiasm of the villagers, but also created a united and uplifting collectivist atmosphere, while further guaranteeing the quality of the tourism products. In addition, the business management model in the form of associations is also an important factor in promoting collective economic development and achieving universal wealth for villagers. Yuanjia Village's more than 30 industries have formed dozens of associations, and the ownership structure and property rights structure are very clear, which also determines the operation and rational distribution of capital. With the expansion of business scale, Yuanjia Village

Figure 6-11　Scenes in Yuanjia Village
(Source: Urban Planning Society of China)

Figure 6-12　Display Multi-party Cooperation Mechanism in Yuanjia Village
(Source: Urban Planning Society of China)

introduced professional management companies and professional managers in a timely manner to participate in management under the leadership of two village committees, ensuring that modern management concepts can adapt to the actual development of rural areas.

6.5.3　Cultural Protection: The Practice of Social Forces Intervening in Rural Heritage Protection

In 2018, the China Yanping Rural Art Season project was launched in Jiulong Village, Jukou Township, Yanping District, Nanping City, Fujian Province, China. The village chief of Jiulong Village invited the project to launch in Jiulong Village from the summer of 2018 by the Shanghai Ruan Yisan Heritage Foundation ("RHF"). A large number of historical buildings and complete traditional village landscapes remain in Jiulong Village. Through exchanges with the villagers and the local government, and taking into account the fact that the vast majority of villagers have moved to the city and the elderly have to stay in the village, the RHF seeks a combination of awakening the awareness of local villagers to protect and develop the countryside and attracting urban residents to the countryside. The method ultimately chose to use the art season to achieve this goal, and persuaded Jiulong Village and the government to participate in the project plan and invest their resources.

In the preparatory process, the Foundation invited nearly 30 artists who are keen to explore the cultural heritage protection and future development of the village to create art in the village. The village elites organized their own assistance to the art season, and 100 local village volunteers participated in the pre-production and opening of the art works. Local craftsmen were invited to complete the works of art with the artists, and more than 10 persons returned home from the cities to open folk houses and restaurants within their own houses. As the event progressed, the villagers began to understand the importance of their traditional earth and wood-structured houses, became proud of their own villages, and began to protect their rural heritage. The opening of the village has attracted many people and organizations, such as volunteers, nature schools and business investment. Jiulong Village has been silent for several years. There is no children's laughter and no hope. Now the village has been awakened by artists, art seasons and people from outside the village. On the opening day of the art season, nearly 5,000 people came to Jiulong Village. After the opening ceremony, there were about 50-100 visitors every day. The most important thing beside the data gathered was the fact that villagers and village elites started thinking about and discussing the future of the previously forgotten and abandoned countryside.

The project successfully involved the local community in the planning, organization and implementation of the art season of the past year. In the art season project, in addition to funding, the local government also played a very important role. The artists invited by the RHF are the most vivid part of the project. Their passion and hard work really touched the local villagers and they began to think about the value of traditional villages and their lives. The art season opened the countryside to the outside world and attracted people from cities such as Shanghai, Fuzhou, Xiamen, France, the Netherlands, the United Kingdom, and the United States. Because of art and heritage, all villagers began to discuss each other's future. This project was awarded the Excellence Award by the International National Trusts Organisation (INTO) (UK) in 2019.

Figure 6-13   Traditional Fujian Brick Houses, Jiulong Village, Yanping, China
(Source: Ruan Yisan Heritage Foundation (RHF), Shanghai, China)

Figure 6-14 Local Villagers Clean the Public Space in the Village Before the Opening Ceremony
(Source: Ruan Yisan Heritage Foundation (RHF), Shanghai, China)

Figure 6-15 The First Bookstore in Jiulong Village Opens on the Opening Day of the Art Season
(Source: Ruan Yisan Heritage Foundation (RHF), Shanghai, China)

Figure 6-16 Artist Zeng Huanguang Uses Local Bamboo and Cloth to Work with Local Craftsmen to Complete the Artwork
(Source: Ruan Yisan Heritage Foundation (RHF), Shanghai, China)

Figure 6-17 Regeneration of the Old Houses - The First Guest house in Jiulong Village
(Source: Ruan Yisan Heritage Foundation (RHF), Shanghai, China)

Figure 6-18 The Village Hall built in 1952, Re-activated as a Venue for Public Events After Having Been Abandoned Nearly 20 years
(Source: Ruan Yisan Heritage Foundation (RHF), Shanghai, China)

6.5.4 Joint Creation: Rural Governance and Construction

Rural governance is the foundation of rural construction and development. The role of rural governance has been constantly emphasized in the rural revitalization strategy and its implementation. In this context, Beijing University of Civil Engineering and Architecture carried out a series of exploration work in Tuguan Village, Jingyang Town, Datong County, and Qinghai Province called "joint creation".

Tuguan Village is located in the southwest of Datong County, 25 kilometers away from Xining City and 15 kilometers away from Datong County. The village is a village for the Tu ethnic group with a long history of more than 500 years. In the process of village planning and implementation, the planners found that the villagers' mentality in waiting for direction and resources from higher-level government was one that had been deeply installed for a long time as the government had dominated many projects, making decisions with less participation and decision-making by the public, which resulted in the village's insufficient development momentum. In order to change this dilemma, the planners practiced "joint creation" through more than a year of exploration. In so doing, they summed up a working system featuring "transforming ideas, changing working methods", "organizing and mobilizing, bringing together multiple parties", and "establishing systems and guaranteeing the joint planning, joint construction, joint management, joint evaluation, and sharing".

In the work of changing the mindset, the planners communicated with the villagers and merged with the villagers, becoming non-common "villagers" themselves. The county and town governments changed their roles from "commanders" to "counselors" in order to serve the villagers, and they organized villagers' training to broaden their horizons and upgrade their skills so that the villagers could change their roles from spectators to participants of change. This helped the villagers establish a confidence in jointly creating a new future for the village. At the same time, the governments at both levels strengthened the building of grassroots party organizations and village autonomy organizations and also established a mechanism featuring "villagers acting as the main body and participating in consultation and joint governance," which guaranteed the participation rights and decision-making power of villagers. All the villagers can use the WeChat group or the suggestion box to make suggestions about the village affairs to the villager's committee and grassroots CPC committee, or they can also report their ideas to the villagers' organizations. These autonomous organizations then propose suggestions to the villagers' congress, and the villagers' congress make the decisions. This system has worked successfully in ensuring the villagers' rights to make proposals, discuss matters and make decisions on such matters that concern them.

In terms of "joint planning", all villagers, government personnel, planning and design teams, and social forces will engage in the planning of the village's industrial development, targeted poverty alleviation, public space remediation, and infrastructure. In terms of "joint construction", a variety of concrete joint construction methods (such as villagers investing in labor and getting paid for specific contributions)are formed for the different types of construction projects in the village. In terms of "joint management", a tri-party joint management and supervision mechanism consisting of the villagers' supervision committee, government personnel, and the joint creation technical team has been formed for the village financial affairs and construction projects. In terms of "joint evaluation", a joint evaluation mechanism for the village environment consisting of students and adult villagers has been adopted, and the mechanism will conduct weekly evaluations of the sanitation and cleanliness of interior space, courtyards and public space once a week. The results of the evaluations will be published. The top ten will receive rewards, and the lowest ten will be criticized. The atmosphere in which the villagers consciously care for the environment has been formed by improving the environmental health awareness of the villagers through rewarding villagers active in maintaining the clean village environment. Through the above series of efforts, all villagers will share the results of environmental improvement as well as the results of poverty alleviation and income increases, and thus, a harmonious and happy living atmosphere will be established.

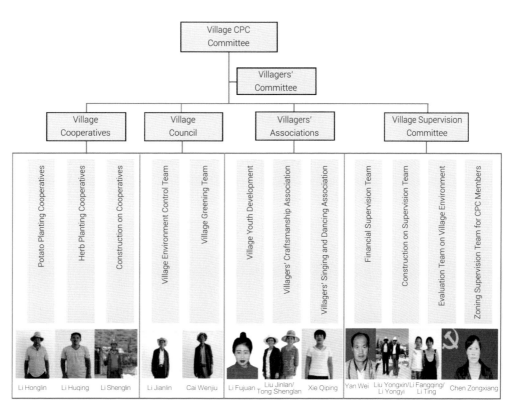

Figure 6-19 "1+4N"Organizational Structure of Tuguan Village
(Source: School of Architecture and Urban Planning, Beijing University of Civil Engineering and Architecture)

Figure 6-20 Craftmanship Training for Villagers to Help Them Get Out of Poverty
(Source: School of Architecture and Urban Planning, Beijing University of Civil Engineering and Architecture)

Appendix I

Appendix 1-1　China Industrial Heritage Protection List (First Batch)

No.	Name	Province and city (county)
001	Couper Dock (now factory area of Guangzhou Huangpu Shipyard)	Guangzhou
002	Jiangnan Machinery Manufacturing Bureau (including Qiuxin Shipyard)	Shanghai
003	Fuzhou Ship-building Bureau (now factory area of Mawei Shipbuilding and within Ship-building Culture Park)	Fuzhou
004	Dagu Dockyard (now the Memorial Hall of Beiyang Navy's Dagu Dockyard Site)	Tianjin
005	Lushun Dock (now factory area of the Liaonan Shipyard [People's Liberation Army Navy 4810 Factory])	Dalian
006	Jinling Machinery Manufacturing Bureau (now Chenguang 1865 Creative Park)	Nanjing
007	Northeast Three Provinces Arsenal (The former site of Shenyang is now Shenyang Liming Aero-Engine Group; the gun office moved to Bei'an to establish Qinghua Tools Factory [626 Factory], now the Qinghua Military Engineering Site Museum)	Shenyang
008	Chongqing Anti-Japanese Warfare Weapon Industrial Site (partially now Chongqing Anti-Japanese Warfare Industrial Site Park)	Chongqing
009	Huangyadong Arsenal (now Huangyadong Arsenal Exhibition Hall)	Licheng County
010	Kailuan Coal Mine (now Kailuan Museum, Kailuan National Mine Park)	Tangshan
011	Zhongxing Coal Mine (now Zhongxing Coal Mine National Mine Park)	Zaozhuang
012	Daye Iron Mine (now Huangshi National Mine Park)	Huangshi
013	Shuikoushan Lead-Zinc mine	Changning
014	Pingxiang Coal Mine (now Memorial Hall of Railway Workers' and Coal Miners' Movement in Anyuan)	Pingxiang
015	Fangzi Coal Mine (Fangzi Coal Mine Heritage Park)	Weifang
016	Fushun Coal Mine (now Fushun Coal Mine Museum)	Fushun City
017	Zhongfu Coal Mine	Jiaozuo
018	Benxi Lake Coal and Iron Company (Fiber Benxi [Xihu] Coal and Iron Industry Site Expo Park)	Benxi
019	Datong Coal Mine (now Jinhuagong Mine National Mine Park)	Datong
020	Fuxin Coal Mine (Haizhou Open-pit Coal Mine National Mine Park)	Fuxin
021	Hanyang Iron Works (Zhang Zhidong and Hanyang Iron Works Museum under construction)	Wuhan
022	Daye Iron Works	Huangshi
023	Anshan Iron and Steel Company	Anshan
024	Capital Steel Corporation (now Shougang Industrial Site Park)	Beijing
025	Changsha Zinc Factory	Changsha
026	Chongqing Steel Works	Chongqing
027	Tangshan Railway Site (proposed China Railway Origin Museum)	Tangshan
028	Chinese Eastern Railway	Heilongjiang Province, Jilin Province, Liaoning Province, Inner Mongolia Autonomous Region
029	Qingdao-Jinan Railway (Qingdao-Jinan Railway Museum [Qingdao-Jinan Railway Jinan Station])	Shandong Province
030	Yunnan-Vietnam Railway (with Yunnan Railway Museum)	Yunnan Province

Appendix I

No.	Name	Province and city (county)
031	Beijing-Zhangjiakou Railway (with Zhan Tianyou Memorial Hall)	Beijing, Zhangjiakou
032	Train Ferry at Xiaguan, Nanjing	Nanjing
033	Baoji-Chengdu Railway	Shaanxi Province, Sichuan Province
034	Bajiaogou-Shibanxi Railway (Jiayang Small Train) (now Jiayang National Mine Park)	Qianwei County
035	Luanhe Iron Bridge	Luanxian County
036	Zhengzhou Yellow River Railway Bridge	Zhengzhou
037	Tianjin Jintang Bridge	Tianjin
038	Shanghai Waibaidu Bridge	Shanghai
039	Jinan Luokou Yellow River Railway Bridge	Jinan
040	Qiantang River Bridge	Hangzhou
041	Wuhan Yangtze River Bridge	Wuhan
042	Nanjing Yangtze River Bridge	Nanjing
043	Chi Hsin Cement Plant (now China Cement Industry Museum)	Tangshan
044	Huaxin Cement Co., Ltd.	Huangshi
045	Conch China Cement	Nanjing
046	Yaohua Glass Factory (now Qinhuangdao Glass Museum)	Qinhuangdao
047	Jiangnan Cement Factory	Nanjing
048	Miaoli Oil Mine (now Taiwan Oil Mine Exhibition Hall)	Miaoli County
049	Yanchang Oilfield	Yanchang County
050	Dushanzi Oilfield, Karamay Oilfield	Karamay
051	Yumen Oilfield (with Yumen Petroleum Museum)	Yumen
052	Daqing Oilfield (with Daqing Oilfield History Exhibition Hall)	Daqing
053	Tangxu Railway Repair Factory (now Tangshan Earthquake Memorial Park, Earthquake Memorial Hall)	Tangshan
054	Dongqing Railway Locomotive Works (Dalian Locomotive Works)	Dalian
055	Feb. 7th Locomotive Works	Beijing
056	Puzhen Rolling Stock Works	Nanjing
057	Tientsin-Pukow Railway Bureau Jinan Machinery Works	Jinan
058	Hip Tung Wo Engineering Works (now Hip Tung Wo Engine Museum)	Guangzhou
059	Zhuzhou Locomotive & Rolling Stock Works	Zhuzhou
060	First Aero-Engine Works of China	Bijie
061	First Auto Works	Changchun
062	First Tractor Works	Luoyang
063	Tianjin Soda Plant (Tianjin Soda Plant Factory History Hall)	Tianjin
064	Yongli Ammonium Sulfate Plant	Nanjing
065	Beijing Coking Plant (Beijing East Industrial Sites Cultural Park)	Beijing
066	Huafeng Paper Mill (Huayuan Idea Factory)	Hangzhou

No.	Name	Province and city (county)
067	Dasheng Cotton Mill (Dasheng Cotton Mill Exhibition Hall)	Nantong
068	Yongtai Silk Factory (China Silk Industry Museum)	Wuxi
069	Yuxiang Cotton Mill	Changsha
070	Dahua Cotton Mill (Dahua 1949, Dahua Industrial heritage Museum)	Xi'an
071	Hangzhou Silk Printing and Dying United Factory (Silian 166 Creative Industrial Park)	Hangzhou
072	Tangshan Ceramic Factory	Tangshan
073	Yuzhou Ceramics Factory (Taoxichuan Cultural and Creative Block)	Jingdezhen
074	Foo Fong Flour Mill	Shanghai
075	Foo Sing No.3 Flour Mill	Shanghai
076	Maoxin Flour Mill	Wuxi
077	Changyu Pioneer Wine Company	Yantai
078	Tsingtao Brewery (Qingdao Beer Museum)	Qingdao
079	Tonghua Wine Co., Ltd.	Tonghua
080	Boyd & Company	Nanjing
081	Shunde Sugar Refinery	Foshan
082	Shanghai Yangshupu Waterworks (Shanghai Waterworks Science & Technology Museum)	Shanghai
083	Hankou Jiji Hydropower Company Zongguan Waterworks	Wuhan
084	Jingshi Tap Water Company Dongzhimen Waterworks (Beijing Tap Water Museum)	Beijing
085	Shanghai East Sewage Treatment Plant	Shanghai
086	Republic of China Capital City Waterworks (Nanjing Tap Water History Exhibition Hall)	Nanjing
087	Shilongba Hydroelectric Power Station	Kunming
088	Xiaguan Power Station (Xiaguan Power Station Site Park)	Nanjing
089	Fengman Hydropower Station	Jilin
090	Shuifeng Hydropower Station	Kuandian County
091	Foziling Reservoir Dam	Huoshan County
092	Sanmenxia Water Control Project	Sanmenxia
093	Chinese Navy Central Radio Station (491 Radio Station)	Beijing
094	Nanjing National Government Central Broadcasting Station	Nanjing
095	Beijing Banknote Printing Plant (541 Plant)	Beijing
096	718 Joint Factory (North China Wireless Joint Equipment Factory) (798 Art Zone)	Beijing
097	Factory 404 (Gansu Mining Area)	Yumen
098	Factory 221 (Qinghai Mining Area) (Nuclear City Memorial Hall)	Haibei Prefecture
099	816 Underground Nuclear Project (816 Underground Nuclear Project Scenic Spot)	Chongqing
100	Jiuquan Satellite Launch Center	Gansu Province, Inner Mongolia Autonomous Region

Appendix 1-2 China Industrial Heritage Protection List (Second Batch)

No.	Name	Province and city (county)
001	Tung-Ka-doo Dock	Shanghai
002	Shanghai Shipyard	Shanghai
003	Dalian Shipyard	Dalian
004	Guangnan Dockyard	Guangzhou
005	China Merchants Steam Navigation Company	Shanghai, Qingdao, Nanjing
006	Qingdao Landing Stage	Qingdao
007	Qinhuangdao Port	Qinhuangdao
008	Dalian Port	Dalian
009	Guangzhou Taigucang Terminal (Baixianke Terminal)	Guangzhou
010	Shanghai Dredging Bureau	Shanghai
011	Dalian Ganjingzi Coal Terminal	Dalian
012	Guia Lighthouse	Macao Special Administrative Region
013	Huaniao Mountain Lighthouse	Shengsi County
014	Eluanbi Lighthouse	Taiwan Province
015	Laotieshan Lighthouse	Dalian
016	Lingao Lighthouse	Lingao County
017	Waglan Lighthouse	Hong Kong Special Administrative Region
018	Naozhou Lighthouse	Zhanjiang
019	Gutzlaff Signal Tower	Shanghai
020	Lanzhou Yellow River Iron Bridge (Zhongshan Bridge)	Lanzhou
021	Tianjin Jiefang Bridge	Tianjin
022	Haizhu Bridge	Guangzhou
023	Imperial Railways of North China (Beijing-Fengtain (Shenyang) Railway)	Beijing, Tianjin, Hebei Province, Liaoning Province
024	Beijing-Hankou Railway	Beijing, Hebei, Henan, Hubei Province
025	Guangzhou-Wuhan Railway	Guangdong Province, Hunan Province, Hubei Province
026	Shijiazhuang-Taiyuan Railway	Hebei Province, Shanxi Province
027	Lanzhou-Shaanxi-Henan-Lianyungang Railway (Lanzhou-Lianyungang Railway)	Jiangsu Province, Anhui Province, Henan Province, Shaanxi Province, Gansu Province
028	Tientsin-Pukow Railway	Tianjin, Hebei Province, Shandong Province, Anhui Province, Jiangsu Province
029	Guangzhou-Kowloon Railway	Guangdong Province, Hong Kong Special Administrative Region
030	Dalian Metropolitan Transportation Co., Ltd.	Dalian
031	Post Office of Qing Dynasty	Tianjin

No.	Name	Province and city (county)
032	Great Northern Telegraph Company	Shanghai, Xiamen
033	Fengtian Machinery Bureau (Shenyang Mint)	Shenyang
034	Central Mint Shanghai, China (State-run 614 Factory)	Shanghai
035	Taiyuan Machinery Bureau (Taiyuan Arsenal)	Taiyuan
036	Gongxian Arsenaol	Gongyi
037	Guantian Central Military Committee Arsenal	Ganzhou
038	State-run 523 Factory (Dalian Jianxin Company)	Dalian
039	Fenxi Machinery Plant	Taiyuan
040	Xikuangshan Mining Administration	Lengshuijiang
041	Guizhou Mercury Mines	Tongren
042	Xihuashan Tungsten Mine	Dayu County
043	Jingxing Mining Bureau (including Jingxing Mine and Zhengfeng Mine)	Shijiazhuang
044	Xiangtan Manganese Mine	Xiangtan
045	Dajishan Tungsten Mine	Quannan County
046	Jinping Phosphate Mine	Lianyungang
047	Shilu Iron Mine	Changjiang County
048	Wenzhou Alum Mine	Cangnan County
049	Keketuohai Mining Bureau (111 Mine)	Fuyun County
050	Jinyinzhai Uranium Mine (711 Mine)	Chenzhou
051	Wangshi'ao Coal Mine	Tongchuan
052	Huili Nickel Mine (901 Mine)	Liangshan Prefecture
053	Qinghai Oilfield	Haixi Prefecture
054	Northwest Steel Works (Taiyuan Iron and Steel Company)	Taiyuan
055	Ma'anshan Iron and Steel Company	Ma'anshan
056	Baotou Iron and Steel Company	Baotou
057	Baiyin Nonferrous Metals Company	Baiyin
058	Panzhihua Iron & Steel Company	Panzhihua
059	Shuicheng Iron and Steel Plant	Liupanshui
060	Shanhaiguan Bridge Factory	Qinhuangdao
061	Shenyang Foundry	Shenyang
062	Taiyuan Heavy Machinery Plant	Taiyuan
063	State-run 331 Factory	Zhuzhou
064	Compagnie de Tramways & D'eclairage de Tientsin	Tianjin
065	Electrical Department of Shanghai Municipal Council (Yangshupu Power Plant)	Shanghai
066	Kailuan Coal Mining Administration Qinghuangdao Power Generation Plant	Qinhuangdao
067	503 Combat-ready Underground Power Generation Plant	Panzhihua

No.	Name	Province and city (county)
068	Riyuetan Daguan Power Station	Nantou County
069	Dongwo Hydropower Station	Luzhou
070	Tianmen River Hydropower Station	Tongzi County
071	Xin'anjiang Hydropower Station	Jiande
072	Liujiaxia Hydropower Station	Yongjing County
073	Qiantang River Seawall Project	Hangzhou, Haining, Haiyan County, Pinghu, Shanghai, Xiaoshan, Shaoxing, Yuyao, Cixi, Ningbo
074	Jinshuizha Floodgate	Wuhan
075	Longyinquan Spring	Dalian
076	Shangli Reservoir	Xiamen
077	Lanzhou Tap Water Company First Waterworks	Lanzhou
078	Yangshupu Workshop of Shanghai Gas Co., Ltd.	Shanghai
079	Acheng Sugar Factory	Harbin
080	Yidu Tea Factory	Yidu
081	A. Lopato & Sons	Harbin
082	British American Tobacco Company	Shanghai, Tianjin, Wuhan, Qingdao, Jinan, Shenyang, Weifang, Xuchang
083	Nanyang Brothers Tobacco Company Limited	Shanghai, Wuhan, Guangzhou
084	Yixin Flour Mill	Wuhu
085	New China Flour Mill (Qianyi Flour Mill Company)	Baoding
086	Shanghai Beer Brewery	Shanghai
087	Shanghai Municipal Council Abattoir	Shanghai
088	Hangzhou First Cotton Printing and Dyeing Factory	Hangzhou
089	Weihui Huaxin Cotton Mill	Weihui
090	Shanghai Boshoku Co., Ltd. Qingdao Factory (Qingdao No.5 State-owned Textile Factory)	Qingdao
091	The Commercial Press	Shanghai, Guangzhou
092	China Alcohol Distillery	Shanghai
093	Dalian Chemical Industry Company	Dalian
094	Tianli Nitrogen Products Factory	Shanghai
095	Yongli Chemical Industry Company Sichuan Factory	Leshan
096	Wuzhou Pine Chemicals Factory	Wuzhou
097	North China Pharmaceutical Factory	Shijiazhuang
098	Qinghai Potash Fertilizer Plant (Qinghai Province Qarhan Potash Fertilizer Plant)	Golmud , Dulan County
099	Central (Hangzhou) Aircraft Manufacturing Company	Hangzhou , Ruili
100	Sino-Italian National Aircraft Works	Nanchang

By National Academy of Innovation Strategy, Urban Planning Society of China.

Appendix II

Appendix2-1 Basic Data of China's 297 Cities at and above Perfecture Level (2016)

Name of Cities	Total area of city's administrative (sq.km)	Total population at year-end (10,000 persons)	Total residents of the Sixth Population Census (10,000 persons)	Area of Built-up district (sq.km)	Gross Regional Product (10,000 yuan)	Per Capita Gross Regional Product (yuan)	Water Coverage Rate(%)	Wastewater Treatment Rate(%)	Per Capita Public Green Space (sq.m)	Domestic Garbage Treatment Rate(%)
Beijing	16411	1363.0	1961.24	1420	256691300	118198	100.00	90.58	13.70	99.84
Tianjin	11917	1044.0	1293.87	1008	178853900	115053	100.00	92.08	16.01	94.16
Shanghai	6341	1450.0	2301.92	999	281786500	116562	100.00	94.29	7.83	100.00
Chongqing	82402	3392.0	2884.62	1351	177405900	57902	97.13	96.75	16.86	99.98
Hebei										
Shijiazhuang	13056	1038.0	1016.38	278	59277293	55177	100.00	96.12	15.77	100.00
Tangshan	13472	760.0	757.73	323	63548675	81239	100.00	97.60	15.28	100.00
Qinhuangdao	7802	298.0	298.76	131	13493526	45280	100.00	96.60	19.22	100.00
Handan	12065	1055.0	917.47	172	33370903	35265	100.00	97.71	18.54	100.00
Xingtai	12433	788.0	710.41	90	19757460	27038	100.00	96.36	11.57	100.00
Baoding	22185	1207.0	1119.44	187	34771269	29992	96.46	90.54	10.18	96.48
Zhangjiakou	36797	470.0	434.55	100	14659911	33142	100.00	94.67	11.93	95.54
Chengde	39493	383.0	347.32	117	14385741	40471	100.00	92.17	24.58	99.52
Cangzhou	14035	780.0	713.41	73	35446800	47425	100.00	99.91	11.00	100.00
Langfang	6382	470.0	435.88	68	27063015	58972	100.00	93.03	13.97	100.00
Hengshui	8815	455.0	434.08	76	14201825	31955	99.63	88.22	12.97	63.38
Shanxi										
Taiyuan	6988	370.0	420.16	340	29556045	68234	100.00	86.89	10.83	100.00
Datong	14176	318.0	331.81	125	10257962	30046	100.00	85.59	11.26	99.75
Yangquan	4570	133.0	136.85	56	6228625	44461	100.00	86.50	11.55	100.00
Changzhi	13896	339.0	333.46	59	12704767	37063	98.54	95.28	12.15	100.00
Jincheng	9425	220.0	227.91	46	10493400	45271	98.98	94.99	12.08	100.00
Shuozhou	10625	163.0	171.49	42	9180640	52010	99.12	97.97	14.14	100.00
Jinzhong	16444	332.0	324.94	77	10911041	32646	100.00	96.96	17.45	100.00
Yuncheng	14183	531.0	513.48	66	12223486	23106	98.49	91.05	14.26	100.00
Xinzhou	25152	308.0	306.75	37	7161357	22747	100.00	95.47	12.81	100.00
Linfen	20275	434.0	431.66	58	12051761	27102	96.62	91.38	12.10	100.00
Lvliang	21239	391.0	372.71	26	9953079	25896	97.83	94.01	13.31	100.00
Inner Mongolia										
Huhhot	17453	241.0	286.66	260	31735900	103235	99.96	94.64	19.69	100.00
Baotou	27768	224.0	265.04	201	38676300	136021	99.55	90.42	13.77	98.15

Appendix II

Name of Cities	Total area of city's administrative (sq.km)	Total population at year-end (10,000 persons)	Total residents of the Sixth Population Census (10,000 persons)	Area of Built-up district (sq.km)	Gross Regional Product (10,000 yuan)	Per Capita Gross Regional Product (yuan)	Water Coverage Rate(%)	Wastewater Treatment Rate(%)	Per Capita Public Green Space (sq.m)	Domestic Garbage Treatment Rate(%)
Wuhai	1669	44.0	53.29	62	5722261	102725	100.00	96.50	20.06	98.60
Chifeng	90021	463.0	434.12	106	19332792	44936	98.60	93.78	17.37	100.00
Tongliao	59329	319.0	313.92	61	19493818	62424	97.90	98.01	21.18	100.00
Ordos	86752	159.0	194.07	117	44179341	215488	99.82	97.27	33.84	97.70
Hulunbeier	252777	259.0	254.93	93	16208500	64140	97.52	99.34	20.30	100.00
Bayannur	66277	175.0	166.99	51	9153800	54480	97.49	98.72	22.56	100.00
Ulanqab	54500	274.0	214.36	60	9388700	44517	97.42	94.68	40.49	96.63
Liaoning										
Shenyang	12860	734.0	810.62	588	55464498	66893	99.91	94.92	11.52	100.00
Dalian	12574	596.0	669.04	396	68101998	97470	99.76	94.73	11.02	100.00
Anshan	9255	346.0	364.59	172	14619713	40532	100.00	86.49	11.09	100.00
Fushun	11272	215.0	213.81	139	8650721	41741	98.62	98.21	10.71	100.00
Benxi	8411	150.0	170.95	109	7667098	44745	99.56	97.08	10.75	86.21
Dandong	14967	238.0	244.47	77	7512352	31223	100.00	87.65	11.07	100.00
Jinzhou	10047	302.0	312.65	112	10328139	33692	100.00	89.79	13.55	100.00
Yingkou	5242	233.0	242.85	189	11562477	47358	100.00	81.59	11.75	71.39
Fuxin	10355	189.0	181.93	77	4078179	22956	98.94	98.18	12.93	100.00
Liaoyang	4788	179.0	185.88	105	6541758	35476	100.00	99.68	10.90	100.00
Panjin	4065	130.0	139.25	75	10071351	70110	98.58	97.21	11.56	71.70
Tieling	12985	300.0	271.77	57	5880423	22178	99.00	99.81	11.97	100.00
Chaoyang	19698	341.0	304.46	57	7165334	24285	91.65	99.09	9.82	100.00
Huludao	10414	280.0	262.35	93	6473518	25347	100.00	89.99	14.72	100.00
Jilin										
Changchun	20594	753.0	767.44	519	59864200	79434	99.81	93.45	17.78	90.27
Jilin	27711	422.0	441.32	189	24535091	57818	98.57	96.14	12.05	100.00
Siping	14382	324.0	338.52	58	11938035	36732	72.21	97.34	8.41	91.10
Liaoyuan	5140	120.0	117.62	46	7652485	63480	95.35	93.89	9.95	100.00
Tonghua	15612	220.0	232.44	54	9475914	42979	94.06	94.73	14.14	95.48
Baishan	17505	122.0	129.61	47	6966243	56411	89.24	86.77	10.21	99.09
Songyuan	21089	278.0	288.01	51	16516898	59413	96.06	96.21	17.79	96.70
Baicheng	25759	193.0	203.24	43	7001392	35892	98.51	80.16	13.00	96.05
Heilongjiang										
Harbin	53100	962.0	1063.60	431	61016096	63445	100.00	92.20	9.21	91.80

Name of Cities	Total area of city's administrative (sq.km)	Total population at year-end (10,000 persons)	Total residents of the Sixth Population Census (10,000 persons)	Area of Built-up district (sq.km)	Gross Regional Product (10,000 yuan)	Per Capita Gross Regional Product (yuan)	Water Coverage Rate(%)	Wastewater Treatment Rate(%)	Per Capita Public Green Space (sq.m)	Domestic Garbage Treatment Rate(%)
Qiqihar	42496	544.0	536.70	140	13253110	25690	98.99	90.93	10.06	68.32
Jixi	22531	181.0	186.22	81	5183793	28647	98.52	74.26	10.82	87.39
Hegang	14679	104.0	105.87	53	2641031	25244	95.78	72.51	14.98	100.00
Shuangyashan	22681	145.0	146.26	58	4373971	29959	98.95	88.37	14.44	86.25
Daqing	21219	276.0	290.45	245	26100031	94690	95.91	96.08	14.97	100.00
Yichun	32800	118.0	114.81	157	2512167	21043	86.38	86.38	23.74	57.38
Jiamusi	32704	238.0	255.21	97	8450332	36878	96.26	85.00	14.17	100.00
Qitaihe	6221	80.0	92.05	68	2166414	26500	98.17	60.84	12.14	98.21
Mudanjiang	38827	259.0	279.87	82	13681181	49618	93.66	100.00	10.61	100.00
Heihe	69345	163.0	167.39	20	4708056	27889	96.97	92.97	13.36	100.00
Suihua	34873	543.0	541.82	45	13163122	24109	98.64	88.90	8.53	100.00
Jiangsu										
Nanjing	6587	663.0	800.37	774	105030200	127264	100.00	95.98	15.34	100.00
Wuxi	4627	486.0	637.44	332	92100200	141258	100.00	97.13	14.91	100.00
Xuzhou	11765	1041.0	857.72	261	58085200	66845	99.81	93.60	15.74	100.00
Changzhou	4373	375.0	459.24	261	57738600	122721	100.00	96.34	14.45	100.00
Suzhou	8657	678.0	1045.99	461	154750900	145556	100.00	95.16	14.71	100.00
Nantong	10549	767.0	728.36	216	67682000	92702	100.00	94.11	18.47	100.00
Lianyungang	7615	534.0	439.35	213	23764800	52987	100.00	87.16	14.66	100.00
Huai'an	10030	568.0	480.17	179	30480000	62446	100.00	93.20	14.01	100.00
Yancheng	16931	831.0	726.22	147	45760800	63278	100.00	90.50	12.75	100.00
Yangzhou	6591	462.0	446.01	149	44493800	99151	100.00	94.42	18.58	100.00
Zhenjiang	3840	272.0	311.41	139	38338400	120603	100.00	94.51	18.97	100.00
Taizhou	5787	508.0	461.89	115	41017800	88330	100.00	91.22	10.69	100.00
Suqian	8524	592.0	471.92	86	23511200	48311	100.00	94.53	15.27	100.00
Zhejiang										
Hangzhou	16596	736.0	870.04	541	113137223	124286	100.00	95.07	14.42	100.00
Ningbo	9816	591.0	760.57	331	86864911	110656	100.00	95.41	11.40	100.00
Wenzhou	12083	818.0	912.21	241	51015586	55779	100.00	92.50	12.73	100.00
Jiaxing	4223	352.0	450.17	101	38621104	83968	99.27	88.32	13.36	100.00
Huzhou	5820	265.0	289.35	106	22843743	77110	100.00	95.43	16.61	100.00
Shaoxing	8279	445.0	491.22	204	47890304	96204	100.00	94.51	13.51	100.00
Jinhua	10942	481.0	536.16	98	36849362	67158	100.00	94.68	11.74	100.00

Name of Cities	Total area of city's administrative (sq.km)	Total population at year-end (10,000 persons)	Total residents of the Sixth Population Census (10,000 persons)	Area of Built-up district (sq.km)	Gross Regional Product (10,000 yuan)	Per Capita Gross Regional Product (yuan)	Water Coverage Rate(%)	Wastewater Treatment Rate(%)	Per Capita Public Green Space (sq.m)	Domestic Garbage Treatment Rate(%)
Quzhou	8845	257.0	212.27	71	12515883	58281	100.00	95.76	14.52	100.00
Zhoushan	1456	97.0	112.13	63	12411989	107463	100.00	95.35	13.09	100.00
Taizhou	9411	600.0	596.88	140	38986594	64287	100.00	93.40	12.88	100.00
Lishui	17298	268.0	211.70	35	12102414	56238	100.00	95.20	11.09	100.00
Anhui										
Hefei	11445	730.0	570.25	460	62743777	80138	99.25	99.71	13.50	100.00
Wuhu	6026	388.0	226.31	172	26994385	73715	100.00	93.56	13.42	100.00
Bengbu	5951	380.0	316.45	145	13858228	41855	100.00	99.51	13.03	100.00
Huainan	5532	389.0	233.39	110	9638395	27990	99.93	97.47	12.60	100.00
Maanshan	4049	229.0	136.63	95	14937617	65833	100.00	99.64	14.98	100.00
Huaibei	2741	217.0	211.43	85	7990337	36427	99.15	97.97	16.72	100.00
Tongling	2991	171.0	72.40	81	9573000	59960	100.00	93.10	17.67	100.00
Anqing	13538	529.0	531.14	90	15311776	33294	100.00	97.37	13.96	100.00
Huangshan	9678	148.0	135.90	67	5768174	41905	100.00	94.54	14.88	100.00
Chuzhou	13516	454.0	393.79	85	14228257	35301	100.00	96.71	14.48	100.00
Fuyang	10118	1062.0	759.99	124	14018589	17642	95.55	94.09	13.95	100.00
Suzhou	9939	654.0	535.29	79	13518116	24270	98.87	98.05	13.41	100.00
Lu'an	15451	587.0	561.17	77	11081469	23298	99.67	98.42	14.84	100.00
Bozhou	8521	647.0	485.07	62	10461044	20611	98.83	94.09	13.39	100.00
Chizhou	8399	162.0	140.25	37	5890196	40919	99.51	93.90	17.08	100.00
Xuancheng	12313	280.0	253.29	55	10578243	40740	99.50	93.94	14.10	100.00
Fujian										
Fuzhou	12675	687.0	711.54	265	61976395	82251	99.99	93.21	14.07	99.00
Xiamen	1699	221.0	353.13	335	37842662	97282	99.81	93.63	11.47	97.75
Putian	4131	350.0	277.85	90	18234281	63313	99.54	85.00	12.70	99.15
Sanming	23095	287.0	250.34	39	18608197	73261	99.86	87.01	14.76	98.60
Quanzhou	11015	730.0	812.85	231	66466294	77784	99.11	95.00	14.20	98.68
Zhangzhou	12554	508.0	481.00	67	31253456	62196	100.00	90.84	14.64	99.70
Nanping	26280	321.0	264.55	41	14577378	55009	100.00	86.90	13.11	95.28
Longyan	19063	314.0	255.95	62	18956670	72354	99.28	89.75	12.51	99.67
Ningde	13247	352.0	282.20	32	16231142	56358	99.22	87.49	15.64	96.49
Jiangxi										
Nanchang	7402	523.0	504.26	317	43549927	81598	98.88	93.50	11.81	99.99

Name of Cities	Total area of city's administrative (sq.km)	Total population at year-end (10,000 persons)	Total residents of the Sixth Population Census (10,000 persons)	Area of Built-up district (sq.km)	Gross Regional Product (10,000 yuan)	Per Capita Gross Regional Product (yuan)	Water Coverage Rate(%)	Wastewater Treatment Rate(%)	Per Capita Public Green Space (sq.m)	Domestic Garbage Treatment Rate(%)
Jingdezhen	5261	169.0	158.75	79	8401484	50989	98.02	68.16	17.18	100.00
Pingxiang	3831	200.0	185.45	52	9982752	52330	100.00	90.13	10.62	100.00
Jiujiang	19798	520.0	472.88	107	20961347	43338	99.30	99.47	17.81	100.00
Xinyu	3178	124.0	113.89	78	10361912	88548	100.00	97.06	18.00	100.00
Yingtan	3560	128.0	112.52	39	6953489	60136	96.11	97.66	15.29	100.00
Ganzhou	39363	971.0	836.84	142	22071959	25761	98.38	85.29	11.45	100.00
Ji'an	25373	535.0	481.03	56	14613721	29772	94.70	91.98	17.09	100.00
Yichun	18669	602.0	541.96	70	17819520	32269	96.97	94.61	15.27	100.00
Fuzhou	18799	401.0	391.23	60	12109070	30259	99.45	93.00	14.69	100.00
Shangrao	22791	782.0	657.97	78	18177664	26996	99.76	78.75	15.49	100.00
Shandong										
Jinan	7998	633.0	681.40	448	65361165	90999	100.00	97.21	11.31	100.00
Qingdao	11282	791.0	871.51	599	100112900	109407	100.00	96.08	18.55	100.00
Zibo	5965	432.0	453.06	271	44120100	94587	100.00	96.40	18.74	100.00
Zaozhuang	4564	413.0	372.91	151	21426335	54984	99.44	95.95	14.98	100.00
Dongying	8243	193.0	203.53	151	34796000	164024	100.00	95.89	22.48	100.00
Yantai	13852	655.0	696.82	330	69256587	98388	97.75	95.84	20.68	100.00
Weifang	16143	901.0	908.62	179	51706000	59275	100.00	95.29	18.07	100.00
Jining	11311	876.0	808.19	199	43018200	51662	100.00	96.10	14.73	100.00
Tai'an	7762	569.0	549.42	155	33167900	59027	100.00	96.65	22.77	100.00
Weihai	5798	256.0	280.48	193	32122000	114220	100.00	96.08	26.09	100.00
Rizhao	5359	300.0	280.10	104	18024900	62357	100.00	95.85	21.23	100.00
Laiwu	2246	129.0	129.85	120	7027600	51533	100.00	94.18	22.59	100.00
Linyi	17191	1141.0	1003.94	208	40267500	38803	100.00	95.35	19.47	100.00
Dezhou	10358	593.0	556.82	154	29329900	50856	100.00	96.71	24.80	100.00
Liaocheng	8984	633.0	578.99	101	28591800	47624	99.42	95.19	12.96	100.00
Binzhou	9660	392.0	374.85	156	24701013	63745	100.00	95.02	19.53	100.00
Heze	12256	1015.0	828.77	125	25602400	29904	99.34	96.52	11.21	100.00
Henan										
Zhengzhou	7446	827.0	862.71	457	81139666	84114	100.00	99.82	8.43	100.00
Kaifeng	6444	559.0	467.65	129	17551002	38619	93.33	93.50	9.30	100.00
Luoyang	15236	737.0	654.99	216	38201075	56410	98.51	99.94	10.46	95.42
Pingdingshan	7882	568.0	490.47	73	18251414	36708	97.90	99.94	10.32	100.00

Name of Cities	Total area of city's administrative (sq.km)	Total population at year-end (10,000 persons)	Total residents of the Sixth Population Census (10,000 persons)	Area of Built-up district (sq.km)	Gross Regional Product (10,000 yuan)	Per Capita Gross Regional Product (yuan)	Water Coverage Rate(%)	Wastewater Treatment Rate(%)	Per Capita Public Green Space (sq.m)	Domestic Garbage Treatment Rate(%)
Anyang	7384	626.0	517.32	82	20298494	39603	100.00	97.73	11.03	100.00
Hebi	2182	170.0	156.92	64	7717894	47940	96.65	92.48	14.57	100.00
Xinxiang	8666	646.0	570.82	118	21669705	37805	99.22	92.00	11.00	100.00
Jiaozuo	4071	374.0	354.01	113	20950796	59183	99.20	95.00	13.20	97.50
Puyang	4188	433.0	359.87	59	14495555	40059	98.18	93.10	14.32	99.80
Xuchang	4997	510.0	430.75	108	23777133	54522	98.16	90.53	12.84	100.00
Luohe	2617	269.0	254.43	67	10819257	41138	88.08	97.59	14.88	100.00
Sanmenxia	10496	229.0	223.40	49	13258631	58894	93.61	95.80	12.04	96.74
Nanyang	26509	1195.0	1026.37	150	31149653	31010	73.81	98.88	8.04	96.52
Shangqiu	12725	977.0	736.30	63	19891538	27332	67.10	78.79	7.33	100.00
Xinyang	18787	908.0	610.91	94	20378010	31733	98.14	90.07	14.14	100.00
Zhoukou	11961	1259.0	895.38	70	22638615	25682	100.00	93.22	13.56	99.32
Zhumadian	15087	949.0	723.12	80	19729881	28305	93.69	96.66	11.22	95.46
Hubei										
Wuhan	8569	834.0	978.54	458	119126100	111469	100.00	97.41	10.39	100.00
Huangshi	4583	270.0	242.93	79	13055500	53033	100.00	92.69	11.85	100.00
Shiyan	23680	348.0	334.08	107	14291500	42083	96.91	98.77	11.08	100.00
Yichang	21230	394.0	405.97	167	37093600	89978	100.00	93.69	14.59	100.00
Xiangyang	19728	594.0	550.03	190	36945100	65663	100.00	93.00	12.43	100.00
Ezhou	1594	111.0	104.87	64	7978200	74983	100.00	92.52	14.93	100.00
Jingmen	12404	300.0	287.37	63	15210000	52470	100.00	96.14	11.83	100.00
Xiaogan	8910	523.0	481.45	79	15766900	32236	100.00	95.90	9.60	100.00
Jingzhou	14243	646.0	569.17	86	17267500	30305	99.81	92.01	10.45	100.00
Huanggang	17457	747.0	616.21	53	17261700	27373	100.00	97.80	13.90	99.15
Xianning	9861	304.0	246.26	66	11079300	44027	98.32	94.71	14.45	51.43
Suizhou	9636	252.0	216.22	71	8521800	38801	94.27	97.92	9.49	95.79
Hunan										
Changsha	11816	696.0	704.10	375	93569088	124122	99.85	96.93	10.75	100.00
Zhuzhou	11307	404.0	385.71	142	24884543	62081	100.00	98.02	12.67	100.00
Xiangtan	5008	290.0	275.22	80	18667869	65946	96.12	95.00	9.34	100.00
Hengyang	15303	799.0	714.83	159	28530158	39020	99.82	92.80	10.17	100.00
Shaoyang	20830	830.0	707.17	72	15302577	20987	95.22	88.81	12.41	98.01
Yueyang	14858	571.0	547.61	100	31008720	54832	100.00	94.56	9.45	100.00

Name of Cities	Total area of city's administrative (sq.km)	Total population at year-end (10,000 persons)	Total residents of the Sixth Population Census (10,000 persons)	Area of Built-up district (sq.km)	Gross Regional Product (10,000 yuan)	Per Capita Gross Regional Product (yuan)	Water Coverage Rate(%)	Wastewater Treatment Rate(%)	Per Capita Public Green Space (sq.m)	Domestic Garbage Treatment Rate(%)
Changde	18190	611.0	571.46	93	29538202	50543	96.34	93.98	13.63	100.00
Zhangjiajie	9534	171.0	147.81	34	4930990	32300	98.03	87.12	9.22	100.00
Yiyang	12320	484.0	430.79	76	14931802	33772	95.59	92.99	9.05	100.00
Chenzhou	19654	535.0	458.35	77	22041285	46691	99.22	93.50	12.12	100.00
Yongzhou	22260	645.0	519.43	64	15658072	28744	98.78	90.32	11.10	100.00
Huaihua	27758	523.0	474.17	64	14003368	28515	91.60	88.80	8.11	100.00
Loudi	8109	453.0	378.46	50	14001393	36058	99.16	91.27	9.55	100.00
Guangdong										
Guangzhou	7434	870.0	1270.19	1249	195474420	141933	100.00	94.28	22.09	100.00
Shaoguan	18413	334.0	282.62	102	12183920	41388	93.75	87.12	12.52	100.00
Shenzhen	1997	385.0	1035.84	923	194926012	167411	100.00	97.62	16.45	100.00
Zhuhai	1732	115.0	156.25	141	22263708	134546	100.00	96.29	19.70	100.00
Shantou	2199	559.0	538.93	258	20809729	37390	100.00	90.32	15.19	89.83
Foshan	3798	400.0	719.74	159	86300002	115891	98.97	96.69	13.91	100.00
Jiangmen	9509	394.0	445.07	152	24187806	53374	99.50	92.10	17.78	100.00
Zhanjiang	13263	835.0	699.48	111	25844327	35612	93.03	91.12	13.99	100.00
Maoming	11429	799.0	581.75	128	26367435	43211	100.00	94.33	16.46	100.00
Zhaoqing	14891	444.0	391.65	120	20840190	51178	98.35	89.49	20.39	100.00
Huizhou	11346	364.0	459.84	263	34121671	71605	98.52	97.02	17.85	100.00
Meizhou	15865	551.0	423.85	58	10455668	24032	92.49	96.58	17.00	100.00
Shanwei	4865	362.0	293.55	22	8284882	27351	97.81	91.21	14.08	93.75
Heyuan	15654	373.0	295.02	38	8987162	29205	100.00	92.52	12.61	100.00
Yangjiang	7956	296.0	242.17	64	12707564	50431	100.00	87.90	12.57	100.00
Qingyuan	19036	432.0	369.84	86	13877104	36136	79.98	81.45	10.00	80.60
Dongguan	2460	201.0	822.02	957	68276868	82682	100.00	93.49	22.99	100.00
Zhongshan	1784	161.0	312.13	149	32027780	99471	100.00	96.30	18.41	100.00
Chaozhou	3146	274.0	266.95	78	9768303	36956	84.89	80.96	9.70	100.00
Jieyang	5265	697.0	588.43	131	20068992	33027	86.54	78.29	12.10	96.42
Yunfu	7787	301.0	236.72	29	7783051	31502	99.81	77.87	19.22	100.00
Guangxi										
Nanning	22244	752.0	665.87	310	37033300	52723	96.14	89.51	12.07	99.04
Liuzhou	18597	386.0	375.87	188	24769396	62855	98.36	95.10	13.46	100.00
Guilin	27667	534.0	474.80	102	20548216	41216	97.17	90.05	11.91	100.00

Name of Cities	Total area of city's administrative (sq.km)	Total population at year-end (10,000 persons)	Total residents of the Sixth Population Census (10,000 persons)	Area of Built-up district (sq.km)	Gross Regional Product (10,000 yuan)	Per Capita Gross Regional Product (yuan)	Water Coverage Rate(%)	Wastewater Treatment Rate(%)	Per Capita Public Green Space (sq.m)	Domestic Garbage Treatment Rate(%)
Wuzhou	12588	347.0	288.22	57	11756486	39072	96.17	90.24	11.14	100.00
Beihai	3337	174.0	153.93	76	10066500	61580	97.76	97.20	10.93	100.00
Fangchenggang	6238	97.0	86.69	41	6760383	73188	100.00	87.32	16.01	100.00
Qinzhou	12187	409.0	307.97	95	11020466	34160	99.92	95.88	12.75	100.00
Guigang	10602	555.0	411.88	73	9587564	22230	98.29	99.51	11.78	100.00
Yulin	12835	717.0	548.74	70	15538300	27111	100.00	99.14	10.23	100.00
Baise	36202	417.0	346.68	49	11143094	30881	100.00	87.05	12.15	100.00
Hezhou	11753	243.0	195.41	66	5181900	25499	99.00	89.16	8.54	100.00
Hechi	33476	429.0	336.93	24	6571808	18842	100.00	93.57	10.30	100.00
Laibin	13411	269.0	209.97	43	5891105	26885	99.93	87.37	10.33	100.00
Chongzuo	17332	251.0	199.43	30	7662005	37161	94.87	34.20	12.92	62.98
Hainan										
Haikou	2304	167.0	204.62	147	12576653	56315	98.46	94.99	12.10	100.00
Sanya	1921	58.0	68.54	56	4755567	63273	97.82	68.02	13.02	100.00
Sansha	13	0.0		0.32			65.00	31.43	3.25	100.00
Danzhou	3400	95.0	93.24	35	2577835	28770	99.77	93.94	13.71	100.00
Sichuan										
Chengdu	14335	1399.0	1404.76	837	121702335	76960	94.95	94.30	14.23	100.00
Zigong	4381	327.0	267.89	116	12345637	44481	77.10	94.82	10.18	100.00
Panzhihua	7401	111.0	121.41	76	10146839	82221	83.56	93.26	11.01	100.00
Luzhou	12236	508.0	421.84	136	14819105	34497	95.43	92.00	10.54	100.00
Deyang	5911	392.0	361.58	75	17524542	49835	93.65	92.01	10.70	98.14
Mianyang	20248	545.0	461.39	139	18304207	38202	93.65	92.67	11.52	100.00
Guangyuan	16319	305.0	248.41	60	6600100	25072	98.16	98.69	11.87	97.13
Suining	5322	378.0	325.26	79	10084521	30615	99.14	99.11	10.23	100.00
Neijiang	5385	420.0	370.28	76	12976712	34667	93.37	90.02	10.63	100.00
Leshan	12723	355.0	323.58	76	14065848	43110	97.10	87.77	7.33	99.53
Nanchong	12477	741.0	627.86	120	16514004	25871	98.33	88.00	12.34	100.00
Meishan	7140	350.0	295.05	64	11172317	37227	96.95	85.53	12.17	100.00
Yibin	13271	556.0	447.19	94	16530529	36735	80.98	87.79	9.93	100.00
Guang'an	6339	467.0	320.55	58	10786241	33130	95.60	96.11	21.82	100.00
Dazhou	16588	684.0	546.81	108	14470836	25921	95.39	46.23	18.69	95.53
Yaan	15046	155.0	150.73	34	5453272	35335	99.51	86.20	10.89	98.34

Name of Cities	Total area of city's administrative (sq.km)	Total population at year-end (10,000 persons)	Total residents of the Sixth Population Census (10,000 persons)	Area of Built-up district (sq.km)	Gross Regional Product (10,000 yuan)	Per Capita Gross Regional Product (yuan)	Water Coverage Rate(%)	Wastewater Treatment Rate(%)	Per Capita Public Green Space (sq.m)	Domestic Garbage Treatment Rate(%)
Bazhong	12293	375.0	328.31	48	5446605	16415	85.59	87.23	12.08	98.00
Ziyang	5748	355.0	366.51	49	9434411	37308	99.70	88.67	15.16	100.00
Guizhou										
Guiyang	8043	401.0	432.26	249	31577001	67772	98.83	97.56	16.18	96.00
Liupanshui	9914	340.0	285.13	73	13137000	45325	90.92	72.36	11.11	95.00
Zunyi	30762	802.0	612.71	120	24039400	38709	94.30	97.13	17.41	95.26
Anshun	9267	300.0	229.76	68	7013500	30216	99.38	93.42	20.43	95.14
Bijie	26849	917.0	653.75	55	16257900	24544	97.74	98.65	21.85	95.00
Tongren	18003	441.0	309.32	48	8569700	27366	92.43	86.37	8.33	92.00
Yunnan										
Kunming	21026	560.0	643.22	436	43000780	64156	98.58	94.07	11.06	96.98
Qujing	28905	653.0	585.51	76	17751063	29266	99.53	92.22	8.93	99.96
Yuxi	15233	217.0	230.35	38	13118823	55389	93.31	93.23	11.18	100.00
Baoshan	19637	261.0	250.65	37	6133904	23692	82.64	85.71	10.07	90.08
Zhaotong	22140	609.0	521.35	42	7655307	14040	97.13	81.02	8.02	100.00
Lijiang	20680	122.0	124.48	24	3092899	24116	98.24	94.13	24.91	93.06
Pu'er	45385	251.0	254.29	27	5675443	21685	92.59	90.13	10.54	98.03
Lincang	23620	237.0	242.95	22	5508172	21906	93.50	92.04	11.91	80.00
Tibet										
Lasa	29518	54.0	55.94	72	4249500	64804	58.46	89.50	4.70	91.85
Xigaze	182000	78.0	70.33	29	1877546	23838	96.56	84.82	33.25	87.83
Changdu	110154	74.0	65.75	7			92.59	53.32	3.33	90.50
Linzhi	116175	19.0	19.51	13			100.00	91.82	9.29	91.19
Shannan	79699	35.0	32.90	15	1265300	35038	77.78	93.37	14.02	91.06
Shaanxi										
Xi'an	10106	825.0	846.78	517	62571800	71357	100.00	92.40	11.87	99.70
Tongchuan	3882	84.0	83.44	40	3116070	36803	92.94	91.45	11.84	90.44
Baoji	18117	384.0	371.67	90	19321400	51262	91.73	91.16	12.34	99.70
Xianyang	10189	530.0	509.60	92	23909700	48016	92.72	92.02	15.37	96.90
Weinan	13134	557.0	528.61	75	14886210	27743	98.62	88.96	12.77	95.00
Yan'an	37037	237.0	218.70	36	10829110	48300	84.94	90.78	10.65	96.50
Hanzhong	27246	384.0	341.62	42	11564920	33597	81.45	91.72	13.37	98.50
Yulin	42923	382.0	335.14	64	27730540	81764	86.66	89.76	12.37	93.27

Name of Cities	Total area of city's administrative (sq.km)	Total population at year-end (10,000 persons)	Total residents of the Sixth Population Census (10,000 persons)	Area of Built-up district (sq.km)	Gross Regional Product (10,000 yuan)	Per Capita Gross Regional Product (yuan)	Water Coverage Rate(%)	Wastewater Treatment Rate(%)	Per Capita Public Green Space (sq.m)	Domestic Garbage Treatment Rate(%)
Ankang	23536	304.0	262.99	45	8428616	31770	95.18	90.26	13.29	99.70
Shangluo	19292	253.0	234.17	26	6992980	29574	99.75	81.64	7.06	96.18
Gansu										
Lanzhou	13086	324.0	361.62	247	22642318	61207	97.02	95.44	12.71	100.00
Jiayuguan	2935	21.0	23.19	70	1534089	62641	100.00	91.19	36.96	100.00
Jinchang	8896	46.0	46.41	43	2078152	44202	100.00	95.17	22.86	100.00
Baiyin	21158	182.0	170.88	63	4422085	25813	100.00	94.09	9.51	95.55
Tianshui	14277	371.0	326.25	56	5905136	17800	96.77	95.71	9.89	100.00
Wuwei	33238	191.0	181.51	32	4617272	25396	97.32	99.71	14.96	99.50
Zhangye	41924	131.0	119.95	64	3999436	32729	100.00	90.61	45.17	100.00
Pingliang	11170	234.0	206.80	36	3673000	17486	99.73	90.50	8.35	100.00
Jiuquan	193974	112.0	109.59	55	5779341	51721	100.00	91.56	11.45	100.00
Qingyang	27119	270.0	221.12	25	5978324	26734	100.00	91.53	7.48	97.40
Dingxi	19609	303.0	269.86	25	3310768	11892	98.36	91.12	16.56	100.00
Longnan	27839	288.0	256.77	14	3398884	13805	95.04	74.74	5.71	100.00
Qinghai										
Xining	7660	203.0	220.87	92	12481677	53756	99.99	74.05	12.21	95.36
Haidong	13161	171.0	139.68	34	4227986	28999	99.49	78.05	5.78	98.50
Ningxia										
Yinchuan	9025	184.0	199.31	171	16177071	74288	92.22	95.21	16.64	97.00
Shizuishan	5310	75.0	72.55	103	5135744	64880	99.79	95.72	23.27	97.72
Wuzhong	16758	142.0	127.38	54	4424283	32039	95.45	90.61	20.55	100.00
Guyuan	13047	150.0	122.82	35	2398058	19720	100.00	90.14	10.06	100.00
Zhongwei	17448	121.0	108.08	54	3391289	29549	90.58	96.25	26.13	100.00
Xinjiang										
Urumqi	13788	268.0	311.26	436	24589766	69865	99.96	90.38	11.35	96.34
Karamay	7735	30.0	39.10	75	6209989	137307	100.00	95.32	11.62	99.08
Turpan	70049	63.0	62.29	19	2251000	35891	100.00	95.00	17.60	100.00
Hami	138919	56.0	57.24	41	4036800	65298	99.95	86.13	13.26	100.00

The data above don't inclucle Hong Kong, Macao or Taiwan.

Notes to the Basic Data of China's 297 Cities at and above Prefecture Level (2016)

I. Data Resources

1. Administrative level
2. Total land area of city's administrative region
3. Total population at year-end
4. Area of built-up district
5. Gross regional product
6. Per capita gross regional product

Source of above data: China City Statistical Yearbook 2017. Beijing: Department of Urban Surveys, National Bureau of Statistics of China. Beijing: China Statistics Press. December 2017.

7. Wastewater treatment rate
8. Domestic garbage treatment rate
9. Water coverage rate
10. Per capita public green space

Source of above data: China Urban Construction Statistical Yearbook 2016, see on the official website of Ministry of Housing and Urban-Rural Development of the People's Republic of China, http://www.mohurd.gov.cn/xytj/tjzljsxytjgb/jstjnj/w02018010521542516551482530.xls

II. Data Illumination

1. As of the end of 2016, 657 Chinese cities nationwide were comprised of 4 municipalities directly under the administration of the central government, 15 sub-provincial cities, 278 prefecture-level cities, and 360 county-level cities.

— *China City Statistical Yearbook 2017*, p.3

2. Total land area of city's administrative region refers to the total area of the land and waters within the administrative region.

— *China City Statistical Yearbook 2017*, p. 397

3. Total population at year-end refers to the total population with residence registration at the public security authorities of the city concerned by 24:00PM, December 31 of each year.

— *China City Statistical Yearbook 2017*, p.397

4. Total Residents of the Sixth National Population Census refers to the permanent population in the Sixth National Population Census conducted at zero hour of November 1, 2010 as the reference time, including persons

living in the local townships, towns or sub-districts, with their household registration at the local townships, towns or sub-districts or with pending household registration; persons living in the local townships, towns or sub-districts and having left the local townships, towns or sub-districts of their household registration for over 6 months; and persons with household registration in the local townships, towns or sub-districts and having left the the local townships, towns or sub-districts for less than 6 months or studying overseas, not including overseas personnel living permanently within the provinces.

— Communiqué of the National Bureau of Statistics of the People's Republic of China on Major Figures of the 2010 Population Census

5. Area of built-up district refers to the contiguous areas within the urban districts which have been actually developed and basically equipped with complete municipal and public facilities.

— *China City Statistical Yearbook 2017*, p.397

6. Gross regional product refers to the final products at market prices produced by all resident units in a region during a certain period of time.

— *China City Statistical Yearbook 2017*, p.398

7. Water coverage rate refers to the ratio of the urban population with access to tap water to the total urban population within the report period. The formula is:

Water Coverage Rate= urban population with access to tap water (including transient urban population) / (urban population + transient urban population) ×100%.

— *Statistical Statements of Cities (County Seats), Villages and Towns Construction* (G.T.Z. [2015] 113)

8. Wastewater treatment rate refers to the ratio of the total sewage treatment volume to the total sewage discharge volume within the report period. The formula is:

Wastewater treatment rate = total sewage treatment volume / total sewage discharge volume ×100%.

— *Statistical Statements of Cities (County Seats), Villages and Towns Construction* (G.T.Z. [2015] 113)

9. Per capita public green space refers to the green space per capita in public space and parks within urban areas at the end of the report period. The formula is:

Per capita public green space = green space in public space and parks within urban areas/ (urban population + transient urban population).

— *Statistical Statements of Cities (County Seats), Villages and Towns Construction* (G.T.Z. [2015] 113)

10. Domestic garbage treatment rate refers to the ratio of the domestic garbage treatment volume to the domestic garbage generation volume within the report period. The formula is:

Domestic garbage treatment rate = domestic garbage treatment volume / domestic garbage generation volume×100%.

— *Statistical Statements of Cities (County Seats), Villages and Towns Construction* (G.T.Z. [2015] 113)

Notes:

1. China City Statistical Yearbook 2017 does not include the land area of administrative regions of the following cities: Changdu, Linzhi and Shannan in the Tibet Autonomous Region, Turpan and Hami in the Xinjiang Uygur Autonomous Region. In the statistical preparation of Basic Data of Chinese Cities in 2016, the above data were taken from the National Administrative Division Information Inquiry Platform of the Ministry of Civil Affairs of the People's Republic of China.

2. China City Statistical Yearbook 2017 does not include the area of built-up districts of the following cities: Taiyuan of Shanxi Province, Shenyang and Dandong of Liaoning Province, Heyuan and Chaozhou of Guangdong Province, Sansha and Danzhou of Hainan Province, Shigatse, Changdu, Linzhi and Shannan of Tibet Autonomous Region, Haidong of Qinghai Province, Turpan and Hami of Xinjiang Uygur Autonomous Region. In the statistical preparation of Basic Data of Chinese Cities in 2016, the above data were taken from China Urban Construction Statistical Yearbook 2016.

3. China City Statistical Yearbook 2017 does not include the gross regional product and per capita gross regional product of the following cities: Sansha of Hainan Province, Changdu, Linzhi, and Shannan of Tibet Autonomous Region, Turpan and Hami of Xinjiang Uygur Autonomous Region. In the statistical preparation of Basic Data of Chinese Cities in 2016, the data on Shannan of the Tibet Autonomous Region were taken from the Statistical Communiqué of National Economic and Social Development of Shannan in 2016, the data on Turpan of Xinjiang Uygur Autonomous Region were taken from the Statistical Communiqué of National Economic and Social Development of Turpan in 2016, and the data on Hami of Xinjiang Uygur Autonomous Region were taken from the Statistical Communiqué of National Economic and Social Development of Hami in 2016.

4. Due to statistical errors, the per capita gross regional product of Qinhuangdao of Hebei Province in 2016 (73,755 yuan) was higher than the per capita gross regional product of the municipal districts in that year (56,805 yuan), and was significantly higher than the city's per capita gross regional product in 2015 (40,746 yuan) according to China City Statistical Yearbook 2017. In the statistical preparation of the Basic Data of Chinese Cities in 2016, the per capita gross regional product of Qinhuangdao was obtained by dividing the gross regional product by the total population at the end of the year (45,280 yuan).

(Data collected and collated by: Mao Qizhi, Professor of Tsinghua University)

AUTHORS

CHAPTER I

Yin Zhi, Vice Chairman, Urban Planning Society of China, Executive Vice President, Professor, Institute for China Sustainable Urbanization Research, Tsinghua University

Lu Qingqiang, Deputy Chief Planner, Director of Research Center for Master Planning, Senior Engineer, Beijing Tsinghua Tongheng Urban Planning and Design Institute

Hu Ming, Urban Planner, Beijing Tsinghua Tongheng Urban Planning and Design Institute

Long Maoqian, Urban Planner, Beijing Tsinghua Tongheng Urban Planning and Design Institute

Yang Gaihui, Urban Planner, Beijing Tsinghua Tongheng Urban Planning and Design Institute

Liu Tingting, Urban Planner, Beijing Tsinghua Tongheng Urban Planning and Design Institute

CHAPTER II

Lin Jian, Dean, Professor, Department of Urban and Regional Planning, College of Urban and Environmental Sciences, Peking University

Liu Shiyi, Postdoctoral Research Fellow, College of Urban and Environmental Sciences, Peking University

Ye Zijun, PhD candidate, College of Urban and Environmental Sciences, Peking University

Wu Ting, Master's degree candidate, College of Urban and Environmental Sciences, Peking University

CHAPTER III

Shao Yisheng, Academician, International Eurasian Academy of Sciences, Research Fellow, China Academy of Urban Planning and Design

Zhang Zhiguo, Vice President, Research Associate, Urban & Rural Water Research Institute, China Academy of Urban Planning and Design

Ma Lin, Deputy Director-General, Professorate Senior Engineer, China Academy of Urban Planning and Design

Zhou Changqing, Director-General, Research Fellow, Research Institute for Water Development, Urban & Rural Water Research Institute, China Academy of Urban Planning and Design

Bai Jing, Research Associate, CAUPD Beijing Planning & Design Consultants Co.

An Yumin, Assistant Engineer, China Academy of Urban Planning and Design

CHAPTER IV

Zhang Quan, Vice Chairman, Research Fellow-level Senior Engineer, Urban Planning Society of China

Ye Xingping, Deputy Chief Engineer, Research Fellow-level Senior Engineer, Urbanization and Urban Rural Planning Research Center of Jiangsu

Chen Guowei, Senior Engineer, Urbanization and Urban Rural Planning Research Center of Jiangsu

Wan Zhen, Assistant Engineer, Urbanization and Urban Rural Planning Research Center of Jiangsu

Li Linyang, Assistant Engineer, Urbanization and Urban Rural Planning Research Center of Jiangsu

CHAPTER V

Shi Weiliang, Vice Chairman, Urban Planning Society of China, President, Professorate Senior Engineer, Beijing Municipal Institute of City Planning & Design

Shi Xiaodong, President, Professorate Senior Engineer, Beijing Municipal Institute of City Planning & Design

Liao Zhengxin, Deputy Director-General, Professorate Senior Engineer, Urban Design Department, Beijing Municipal Institute of City Planning & Design

Ye Nan, Deputy Director-General, Professorate Senior Engineer, Urban Design Department, Beijing Municipal Institute of City Planning & Design

Guo Jing, Senior Engineer, Beijing Municipal Institute of City Planning & Design

Xin Ping, Planner, Beijing Municipal Institute of City Planning & Design

CHAPTER VI

Zhang Shangwu, Director, Chairman of Rural Planning and Development Committee, Urban Planning Society of China, Vice Dean, Professor, College of Architectural and Urban Planning, Tongji University

Xi Hui, Senior Researcher, Senior Engineer, China Rural Planning and Development Research Center, Shanghai Tongji Urban Planning and Design Institute Co., Ltd.

Zhang Li, Secretary-General, Small Towns Planning Committee, Urban Planning Society of China, Associate Professor, College of Architectural and Urban Planning, Tongji University

Yang Ben, Research Fellow, China Rural Planning and Development Research Center, Shanghai Tongji Urban Planning and Design Institute Co., Ltd.

Zou Haiyan, Research Fellow, China Rural Planning and Development Research Center, Shanghai Tongji Urban Planning and Design Institute Co., Ltd.